Losing Sight of the Shore

Losing Sight of the Shore

Scotland's Medical Explorers
1815–1915

Wendell R. McConnaha, Ph.D.

Whittles Publishing

Whittles Publishing Ltd,
Dunbeath,
Caithness, KW6 6EG,
Scotland, UK

www.whittlespublishing.com

© 2025 Wendell R. McConnaha

ISBN 978-184995-594-2

Printed by 4edge Limited, UK

Contents

Conclusion 160

Author's note

There are three points I'd like to clarify. First, while throughout this book I use the term "discovered" when discussing locations, topographical features, or plants and animals identified or collected by the medical explorers, I am fully aware that the indigenous people in the relevant locales would have known of these "discoveries" for generations. These references are meant, then, to denote the first recorded sightings by someone from a Western culture, in this case European.

Second, the focus of this book is to illustrate the role that Scottish doctors played in leading the British Imperial Century. I am aware that many aspects of Britain's empire building have been criticized for often being carried out at the expense of the local populations. While I acknowledge this fact I do not discuss it here, as that is not the focus of this work. I would say, however, that to the best of my knowledge all of the medical explorers discussed here benefited from their acceptance of local cultures, customs, and languages, and also from their friendships with the indigenous people with whom they worked.

Third, because the terms are often incorrectly substituted for each other, it is important to understand the differences between England, Great Britain, the United Kingdom, and the British Isles. In 1567 England was ruled by Elizabeth I and Scotland by James VI. Elizabeth died without heirs and her closest Protestant relative was James. The Union of Crowns declared Scotland and England would remain separate countries but would be ruled by the same monarch and James VI of Scotland also became James I of England. This continued until 1707, when the Acts of Union merged Scotland, England and Wales into Great Britain. The islands of Great Britain and Ireland, along with over 1,000 smaller surrounding islands form the British Isles. Northern Ireland was created in 1921 when Ireland was partitioned by the Government of Ireland Act. Northern Ireland joined England, Scotland and Wales in the United Kingdom.

WRM

Dedicated to Judy my wife and co-author

Foreword

As a thoroughly out-of-place 22- or 23-year-old, I tried to look inconspicuous in my new suit, sitting next to the deputy headmaster, at my first—and only—meeting of the Old Boys' association from my school. I was sitting next to the deputy head because, despite the 30-year age gap between us, he and I were the two youngest people in the room. The crowd listened attentively as the after-dinner speaker's voice rose as he built to a conclusion, "People criticize the British Empire because, they say, everywhere we went we had a sword or a gun in our right hand." He continued, "But they never mention that we always carried a Bible in our other hand." It was the sort of thing his audience at one of London's grandest gentlemen's clubs wanted to hear.

But there is another side to this century of exploration, and Dr. Wendell McConnaha investigates these elements in this publication, *Losing Sight of the Shore.* I first met him as part of my work involving his first book, *The King of Lokoja,* a biography of a remarkable Scottish physician/explorer in the 19th century. With this, his latest, work he expands his study of Scottish medical explorers.

Wendell does not engage with the rights and wrongs of Britain's imperial past. Rather, he looks at why the Scottish medical explorers emerged, and how their emergence was, and still is, reflected in the intricate fabric of Scottish history and culture. This carefully researched, comprehensive and captivating account puts the Scottish medical explorers into their historical and cultural context, and illuminates several of the most remarkable individuals who played key roles in this story.

Along with their Bible and sword these men would carry a stethoscope, a microscope, plant-collecting kits, and surveying equipment. Although in the latter years of this account the globe had been "discovered" and charted, these men still sought science and understanding. This book considers how Scottish politics, scholarship, heritage, mindset and innovative education led so many Edinburgh doctors to explore the vast British, or—as some have even called it—Scottish Empire. Dr. McConnaha proposes that "understanding the evolution and progression of Scottish medical education is a critical component in comprehending why Scotland in general, and Edinburgh in particular, were so well positioned to produce the medical explorers who would lead the British Imperial Century."

It is a bold claim, but it is an argument that he constructs with care, and that he illustrates through his accounts of the lives and experiences of a selection of medically trained explorers who fitted this model. And if any of them are heroes of yours—as the Orcadian explorer John Rae is of mine—be reassured. Seeing them from this new perspective, as exemplars of a much bigger trend, will not in any way diminish them. Rather, it will enhance your understanding and appreciation of these remarkable men.

Huw Williams,
BBC Orkney

Acknowledgements

In formulating my hypothesis regarding the medical explorers of Edinburgh, as well as then pursuing the information that could prove or disprove my theory, I have relied on a great number of individuals. I would be remiss if I were not to give credit to those who made my search possible and to those who contributed to the final creation. I am only hopeful that I don't inadvertently omit someone who richly deserves credit. Should that occur, my apologies in advance.

I will begin by thanking Huw Williams, BBC Orkney. Shortly after completing an earlier biography, I was contacted by Huw to complete an on-air interview related to the publication of my upcoming book tour in Orkney. During my time in Kirkwall, I further connected with Huw, who I found to be a wealth of knowledge on Scotland in general and Orkney in particular. I am indebted to him for the hours of conversation, for his directing me to additional sources of information, and for his agreeing to write the foreword to this book. Also, in Orkney I was introduced to Andrew Lind and Andrew Appleby. Dr. Lind is a lecturer in Northern Studies at the University of the Highlands and Islands, and Andrew Appleby is with the John Rae Society in Kirkwall. Both have made valuable contributions to this work. Finally, my Orkney connections include Lucy Gibbon and Sarah Maclean from the Orkney Library and Archives, and Ellen Pesci, Curator for the Orkney Museum, and I thank each of them for their assistance.

The information gathered regarding the Royal Navy in general, and the Royal Hospital Haslar in particular, relied on materials provided by my friend and colleague Eric Birbeck. A chance introduction over 20 years ago led to a close friendship and a collaborative research partner. Eric Birbeck, MVO, served in the Royal Medical Service for 32 years, including overseeing the medical center onboard Her Majesty's Royal Yacht *Britannia* for five years. In 1996 he transferred into the civil service at the Royal Hospital Haslar. It was shortly after his transfer that he invited me to visit him at the hospital. We spent time together, and I discovered that no one knows the 260+ years of history associated with Haslar better than Eric. He has since become a founding member of the Haslar Heritage Group. The information provided about Haslar, the Royal Navy, and several of the pictures found in the book I owe to Eric. I thank him for his ongoing support.

I first met Julie Carrington, a research librarian at the Royal Geographical Society (RGS), while researching an earlier work on William Balfour Baikie. The RGS had played a significant role in most of expeditions discussed in *Losing Sight of the Shore* and Julie proved extremely helpful in directing my research, providing the needed documentation, and linking me with the medical explorers' personal and professional correspondence. I found her extremely knowledgeable and a first-rate researcher; she is someone whom I have come to think of as a friend and colleague.

The Royal Botanic Gardens Kew figure prominently throughout this book. Many of the facts related to the various individuals profiled were only available in their personal correspondence, which is housed at Kew. My efforts in locating the various pieces of correspondence and illustrations were made possible by Kay Pennick, library assistant; Kat Harrington, assistant archivist; Kiri Ross-Jones, senior archivist and records manager; and Trishya Long, illustrations team member. I am indebted to each of them for their ongoing assistance.

In the early days, as I was formulating what would emerge as the premise of this book, I turned to Lisa Rosner, a professor of history at Stockton College in New Jersey. Her knowledge of the history of medicine in general, and the University of Edinburgh's School of Medicine in particular, greatly assisted my research efforts. When researching the transitions involving the various Scottish monarchs and the relationship between England and Scotland throughout this process, I was in contact with Robert Bucholz, Professor of History at Loyola University Chicago, and Karin Bowie, Senior Lecturer in Scottish history at the University of Glasgow. My thanks to both of them for their assistance.

As with my first book, *The King of Lokoja*, I turned to my good friend Walter Vatter, who spent decades in the New York publishing field and has continuously given me valuable advice and support. He provides an insight that only one who has spent his life working with authors can deliver, and I am indebted to him.

Next, three extremely important acknowledgements. First, I would like to thank artist Carole Beach. with whom I share a studio in Chicago. She and her husband Bill are close friends; we travel together, share the same interests, and use each other as sounding boards for our respective projects. Second, Caroline Petherick was selected by the publisher to serve as the copy editor. I have found her to be knowledgeable, professional and amazing to work with. Her edits which have been incorporated into the final draft have made the book a far better read. Finally, my wife Judy has been involved throughout this process. She assisted with the research, read, and edited countless series of first through final drafts. Her changes, additions and edits all became part of the final work, and she is in every sense my co-author.

I also used a number of sources provided through the writings of some very talented authors. For the chapter on Archibald Menzies, I relied broadly on his *Alaska Travel Journal of Archibald Menzies 1793–1794* (1993), along with two extremely well-written books. The first is *Seeds of Beauty: Scottish Plant Explorers* (2008) by Ann Lindsay. The second is *Monkey Puzzle Man: Archibald Menzies Plant Hunter* (2008) by James McCarthy. The chapter related to John Kirk relied heavily on *The Physician and the Slave Trade: John Kirk, the Livingstone Expeditions, and the Crusade Against Slavery in East Africa* (1999) by Daniel Liebowitz. I also used information found in *The White Nile* (1960) by Alan Moorehead, and *The Exploitation of East Africa* (1967) by Sir Reginald Coupland. The chapter on John Rae was greatly assisted

Acknowledgements

by Rae's *Arctic Journals of John Rae* (2012) as well as *Fatal Passage: The Story of John Rae: the Arctic Hero Time Forgot* (2001) by Ken McGoogan. For the chapter on Charles Wyville Thomson, I utilized *Endless Novelties of Extraordinary Interest* (2019) by Doug MacDougall, *The Silent Landscape* (2003) by Richard Corfield, *The Voyage of the Challenger* (1972) by Eric Linklater, and *The Depths of the Sea* (1873) by Charles Wyville Thomson. For the chapter on William Bruce Speirs, I relied on a number of first-rate sources, including his own words from *Polar Exploration* (1911), together with *William Speirs Bruce: Polar Explorer and Scottish Nationalist* (2003) by Peter Speak, and *William Speirs Bruce: Forgotten Polar Hero* (2018) by Isobel P. Williams and John Dudeney.

I have provided the source and, where required, permission for each of the illustrations, and every effort has been made to ensure the accuracy of names, places, events, and dates, but of course, all responsibility for the content of this book rests with the author alone.

Timeline

50 BCE   The Romans arrive in Britain and find "a basic religious homogeneity" among the Celtic people led by the druids

100   The Celtic religious practices began to display elements of Romanization, resulting in a hybrid Celtic faith

430   As Christianity begins to make inroads into the religious practices of the Celts, the Romans leave Britain

500   By the fifth and sixth centuries ancient Celtic worship has been completely overtaken by the Roman Catholic church

937   Beginning with the Battle of Brunanburh, England and Scotland fight almost continuously for the next 800 years

1295   An alliance, to become known as the Auld Alliance, is created between the kingdoms of Scotland and France

1314   Robert the Bruce's victory over Edward II at the Battle of Bannockburn forces England to acknowledge Scotland as an independent nation

1485   Henry VII, on winning the Battle of Bosworth Field, becomes the first Tudor monarch of England

1505   The Town Council of Edinburgh grants a Charter of Incorporation to the Barber Surgeons of Edinburgh

1530   The Scottish Reformation, with the formation of the Church of Scotland—that is, the Kirk—ends control by the Roman Catholic Church

1560   The Auld Alliance ends with the signing of the Treaty of Edinburgh

1582   The University of Edinburgh is founded under the authority of the Town Council of Edinburgh

1600   The English East India Company receives its Royal Charter from Queen Elizabeth I

1603   Queen Elizabeth I dies leaving no heirs, and the Tudor dynasty ends

1603   With the Union of the Crowns, Scotland's King James VI becomes King of Scotland and England, which begins the Stuart dynasty

1639   The Bishops' Wars begin

1650   Cromwell invades and defeats Scotland

1657   The surgeons end their association with the barbers, then form the Incorporation of the Surgeon Apothecaries under the Town Council of Edinburgh

1670   Sir Robert Sibbald and Andrew Balfour develop a physic garden, which ultimately becomes the Royal Botanic Garden Edinburgh

1670   The Hudson's Bay Company is granted a Royal Charter

1681   The Incorporation of Physicians is granted a Royal Charter by King Charles II, creating the Royal College of Physicians of Edinburgh (RCPE)

1689   William III and Mary II are invited to jointly assume the English throne following the Glorious Revolution

1699   The surgeons of Edinburgh build an "anatomical theatre," which is incorporated into what becomes known as Surgeon's Hall

1701   In the Act of Settlement William III inserts his cousin Sophia of Hanover into England's line of succession

1707   Two Acts of Union, forming England, Scotland, and Wales into a single kingdom, create the Kingdom of Great Britain

1714   Queen Anne, the last of the Stuart rulers, dies, and upon her death George I becomes the first Hanoverian ruler of Great Britain

1726   University of Edinburgh School of Medicine is established

1729   Alexander Monro (primus) establishes the Edinburgh Infirmary for the Sick Poor (known as the Little House) as a clinical site for preparing doctors

1740   The Scottish Enlightenment begins

1753   The British Museum is founded

1754   Archibald Menzies is born

1759   The Royal Botanic Gardens Kew is founded as part of a royal estate set aside as a physic garden

1762   The Royal Hospital Haslar opens

1766   Sir Joseph Banks is elected to the Royal Society and becomes an advisor to King George III at Kew Gardens

1778   A Royal Charter is granted to the Incorporation of the Surgeon Apothecaries, which becomes the Royal College of Surgeons of the City of Edinburgh (RCSE)

1788   Led by Sir Joseph Banks, the Saturday's Club becomes the African Association

1801   The Act of Union creates the United Kingdom of Great Britain and Ireland[*]

1813   John Rae is born

1815   The British Imperial Century begins

1820   The Scottish Enlightenment ends

1820   Sir Joseph Banks and King George III die

1830   Sir Charles Wyville Thomson is born

1831   The Geographical Society of London absorbs the African Association within its membership

1832   Sir John Kirk is born

---

[*] NB As a consequence of the 1921 division of Ireland, it was in 1927 that the UK gained its current title: the United Kingdom of Great Britain and Northern Ireland.

1838   Sir John Richardson arrives at Haslar as chief physician

1841   William Jackson Hooker is appointed director at the Royal Botanic Gardens Kew

1842   Archibald Menzies dies

1846   Sir Edward Parry is posted to the Royal Hospital Haslar as captain superintendent

1858   Parliament issues the Government of India Act, which transfers the power of the British East Company to the Crown

1859   Queen Victoria grants the Geographical Society of London a Royal Charter, and it becomes the Royal Geographical Society

1865   Joseph Dalton Hooker succeeds his father as director of Kew Gardens

1867   William Speirs Bruce is born

1869   The Hudson's Bay Company is directed by Parliament to sell its land to the Canadian government, in the Deed of Surrender

1882   Sir Charles Wyville Thomson dies

1893   John Rae dies

1915   The British Imperial Century ends

1921   William Speirs Bruce dies

1922   Sir John Kirk dies

Introduction

While completing a biography of a Scotsman named William Balfour Baikie (1825–1864) I found that he completed his medical studies at the University of Edinburgh, graduated at the top of his class, and joined the Royal Navy as an assistant surgeon. He was a brilliant physician who played a major role in proving that the use of quinine could prevent malaria. Yet surprisingly his major contributions were in the areas of natural history and exploration—not in the practice of medicine. As work on this biography progressed, I consulted with the organizations with whom Baikie had been affiliated: the Royal Navy, the Royal Botanic Gardens Kew, the Royal Hospital Haslar, the British Museum, the Royal Geographical Society, the University of Edinburgh, and others. And some intriguing facts emerged.

When Baikie completed his medical studies in 1846, Great Britain was nearing the middle of what has been designated the British Imperial Century. History is more than just dates and events. But both are important to this story. There were specific events, happening at specific times, that ushered in Britain's century of global greatness. The country that emerged from the Napoleonic Wars of 1793–1815 was very different from Britain before the conflict. This was true in matters political and economic, but in addition a profound feeling of nationalistic pride had emerged.[1] Britain's overwhelming victories over Napoleon's forces at Trafalgar in 1805 and at Waterloo in 1815 had left the country, with its empire, the strongest, richest, and most powerful nation in the world. The Royal Navy possessed more ships than the fleets of the next three maritime powers combined.[2] This meant that while every other land in Europe had been repeatedly overrun by invading armies, Britain alone was free to work on new discoveries without interruption.[3]

During the next 100 years Britain moved forward, concentrating on exploring and expanding its already extensive global empire. Its agents' efforts were extremely successful, and during the British Imperial Century (1815–1915) nearly 10 million square miles of territory and roughly 400 million people were added to the British Empire through a combination of colonization and conquest; by 1820 over 25 percent of the world's population were subjects of the British Empire.[4] As the British acquired territory, there was a need to map and chart the lands hitherto unknown to them. Building trade and commerce required individuals who could assess people, plants, animals, minerals, and all nature of things to determine how they might prove to be of commercial value to Britain. Each of these endeavors required people with writing and communication skills to record all they had seen in such a way that it was useful to the governmental agency, merchant, or scientific body who had sponsored the exploration. Each expedition also required a breed of explorer who had the physical strength, as well as educational capabilities, to accomplish these multiple tasks. I understood that these attributes were clearly present in Baikie—but did his remarkable work stand alone, or were these characteristics shared within a larger group?

In the period from 1815 to 1915 Scotland's population was about one-fifth that of the population of England. Despite this difference in the populations of the two countries, over one-half of the British explorers I could identify from that time were Scottish. But although the rise of the British Empire had coincided with England's merger with Scotland in 1707, and culminated during the British Imperial Century, it would be overreaching to rename it the Scottish Empire, as the author Duncan Bruce indicates others have done. He adds that it is, however, clear that without the participation of the Scots the empire would not have been constructed.[5]

My first question immediately arose. How did a Scottish population that was only 20 percent the size of England's provide such a disproportionate number of men to explore, chart, and collect around the world? Investigating further, I found that in addition to the number of explorers inordinately favoring the Scots, many these explorers were, like Baikie, physicians and surgeons who had received their training at Scottish universities. Lisa Rosner begins *Medical Education in the Age of Improvement: Edinburgh Students and Apprentices 1760–1826* with the following sentence. "Young men became medical students at Edinburgh University in order to set up medical practice." [6] There is no doubt that her assertion is accurate for many of the graduating physicians and surgeons. However, my research showed that a large number of 19th-century doctors who had completed their medical preparation elected to reject the traditional role of physician or surgeon in favor of exploration.

Not all the explorers were Scottish. There was a scattering of Englishmen who studied in Scotland, as opposed to Oxford, Trinity, or Cambridge. Not all the Scottish explorers had attended Scottish universities; there were a few Scotsmen who had elected to study in England or on the Continent. But a significant number of the British explorers during the Imperial Century were Scotsmen who had graduated from one of the Scottish university medical schools—and the University of Edinburgh graduates within this unique exploring group outnumbered those from other Scottish universities by a ratio of almost 9 to 1.

What was it, then, that was unique to the preparation of Scottish doctors in particular, and those graduating from the University of Edinburgh in particular, that resulted in such a dramatic change of vocation? My research on Baikie had created a picture of one man, but it now suggested that an entire group of Scottish doctors had selected a similar career path. Baikie and his fellow medical graduates were part of a newly evolved group of doctors who would not forge a career in the hospitals and medical offices of Scotland and England, but rather would explore territories from the jungles of Africa to the ice of the Arctic, and indeed the entire world.

This begged a new set of questions to be addressed. How and why did this happen? Why did these newly qualified young Scottish doctors choose this path, and why were they more uniquely able to lead this effort than their English counterparts? The information collected from my research identified a sequence of events that was particular to these medical explorers that seems to have placed them in a position to assume leadership at exactly the right time. The first of those events was the evolution of the connection between Scotland and England. The early relationship between the two countries was stormy at best, and for hundreds of years the two nations were involved in almost constant conflict, ranging from multiple border raids to full-on war. However, the Union of Crowns in 1603 saw James VI of Scotland being crowned as James I of England as well, and with this act, warfare between the two countries largely came to an end. This union left Scotland and England as separate nations, but from that point onward they would be ruled by a single monarch. Then a century later, in 1707, the Acts of Union merged those two countries, along with Wales, into what would be known as the Kingdom of Great Britain. Suddenly, options such as the military, working within joint-stock companies,* and individuals accessing positions within the scientific and cultural institutions located south of Scotland's border—options previously available only to the English—now were open to the Scots as well.

The Acts of Union also resulted in the English and Scottish parliaments being merged into a single entity located in London. The men who had been involved in the governmental functions in Scotland and who were considered their nation's elite were forced by the Union to leave their country and, some would say, remove a large portion of Scottish nationalism in the process. However, the Acts specifically spared Scotland's Church, its legal profession, and its educational system, guaranteeing that they would remain uniquely Scottish. This meant that the leaders within the Church, the law, and the universities, emerging as the new top stratum of society, created the second event along this continuum: the Scottish Enlightenment. This was a time in which Scottish intellectual life flourished as never before. The great French philosopher Voltaire described Edinburgh during this period as an intellectual center to rival Paris, and Scotland was the envy of world. The Scottish Enlightenment produced a socially and intellectually advanced population. Improved education was one of the primary outcomes, especially at the university level, where the Scottish system was completely transformed.

This led to the third unique element in the emergence of the Scottish medical explorers. Although there was improvement throughout the Scottish university system before and during the Enlightenment, among the most impactful accomplishments was the creation of a new methodology for preparing doctors. Scottish medical schools began to use highly accomplished professors to conduct courses in their specialties. In addition, the advent of teaching hospitals with clinical practice allowed medical students, for the first time, to undertake real-life practical experience with patients. Those who wanted to gain additional knowledge could enroll in the city of Edinburgh's extensive extra-mural offerings and take courses outside their compulsory studies from private lecturers. Of special importance to the development of these future explorers and the focus with which they viewed the world was the incorporation of classes in the humanities in the medical curriculum. This, along with

---

* Private trading companies legally empowered to perform the essential functions of government and to pay themselves out of the profits.

an infusion of biological, physical, and natural science courses, in addition to the traditional study of anatomy and surgical techniques, broadened and expanded the students' views not only of medicine but of the entire natural world. As a result, the medical network within the city of Edinburgh produced a graduate quite different than those matriculating from English or Irish universities. The Scots were trained to be intellectually curious and analytical. They were broadly educated in the sciences, and when they had completed their studies they were both ambitious and adventurous.

The final variable in the development of these doctors is that the British now required a new type of explorer. The motivation for exploring during the Age of Discovery (1400–1650) or the First (1497–1763) and Second (1783–1815) British Empires had been to find new lands, establish trade routes, and create colonies. Vast numbers of people and large numbers of colonial territories continued to be added during the British Imperial Century, and although trade and profit were still significant considerations, new and never before seen focuses such as economic botany and voyages simply for the sake of science replaced the earlier motivations or worked in conjunction with them. The British Imperial Century was marked by great changes in what constituted the known world. Entire continents, including Australia, Africa, and Antarctica, were being mapped and charted, most of them for the first time. New species of plants and animals were being "discovered" in record numbers, and museum and botanic collections were doubling and then redoubling in size. So, the nature of exploration was changing, and this transformation required a new type of explorer. These men had to be willing to travel, be physically strong, and have a curious mind, with the training and knowledge to see what was new, and the intellect and ability to explain their finds and their impact to others upon their return home. The interest in this new approach began to grow, and the study and understanding of the natural world became integrated within these exploring missions alongside the existing efforts related to the expansion of commerce within the British Empire.

The convergence of these occurrences had a major impact on the psyche of the Scots. These events culminating at this particular time in Scotland's history provided the men of Scotland a trajectory that was quite different than that of their English counterparts. But I would suggest there were additional factors—factors that epitomize the aspects of Scottishness—that drove these individuals forward. These were men of the working and middle classes who were used to hardship and sacrifice. They could face the physical challenges of an exploring life, but also had the intellectual preparation to understand and record what they had learned. They had grown up with a familiarity with the sea and the tradition of the sea as a living and a way of life. Many had the willingness to leave their families and the familiarity of home in pursuit of a better future. These men had little to keep them in Scotland, and so they readily accepted the challenges when others would not.

William Balfour Baikie was a doctor, explorer, and naturalist. Although he was exceptional, my research verified that he was not unique. Rather, he was one of many Scotsmen who had benefited from their country's merger with England, the intellectual stimulation and advancements in education provided by the Scottish Enlightenment, and the transformation in the methodology that the city of Edinburgh and the university's medical school used to prepare its doctors and surgeons. Although *Losing Sight of the Shore* profiles five of the medical explorers in detail, this is not a biography. It is a synthesis of the events preceding and occurring within the British Imperial Century, and an account of how these factors converged to create the characteristics seen in the Scottish medical explorers.

I have selected explorers whose stories I feel validate my argument. The appendix lists nearly 70 men who fit the characteristics identified, and over 30 regions, countries, and continents are listed among their exploring destinations, and any one of these individuals could have been selected to illustrate the soundness of my thesis. Having considered each of these explorers I selected five individuals using the following criteria.

First, although not a great deal has been written by, or about, these men, I considered primary sources essential. Most of these explorers kept journals and other original documents and correspondence that could be used to tell their story.

Second, their lives and periods of exploration needed to cover the full span of the British Imperial Century. The earliest explorer profiled was born in 1754 and the last lived until 1922. Their periods of exploration range from 1795 to 1915.

Third, their reasons for exploring needed to cover the gamut of the changing nature of exploration during that time period.

In the end, in order to substantiate my theory, I selected Archibald Menzies, Sir John Kirk, John Rae, Sir Charles Wyville Thomson, and William Speirs Bruce. Each one was molded by his Scottish heritage, the benefits provided by the impact of the Scottish Enlightenment, and the extraordinary training as physician and surgeon that he had received at Edinburgh. Each benefited from the employment options created by the union of England and Scotland and the changing focus of exploration during the British Imperial Century.

Although their interests in exploring were formed through the same set of variables, the emphasis of their journeys varied widely. Menzies traveled as a naturalist, collecting plants on four continents. Kirk made the eradication of slavery in Africa his life's work. Rae emphasized the economic aspects of exploration, attempting to locate the Northwest Passage to further Britain's commercial success in North America. Thomson conceptualized the science of oceanography, exploring the oceans of the world solely for the sake of science. Speirs spent the greater portion of his life investigating, mapping, and charting the Arctic and Antarctica with the single goal of advancing geographic and other scientific knowledge. The range of motivations and reasons for exploration, and the individual accomplishments of each of these great men, illustrates the vast diversity of exploration during this period. The five individuals selected represent the key elements identified and required to be a 19th-century medical explorer. Their stories illustrate the diversity of the doctors' explorations and the breadth of their accomplishments. This work is an intellectual history which aims to understand and explain the ideology and philosophy of a group of Edinburgh-trained physicians and surgeons, and how their interaction within the political, cultural, intellectual, and social context of the time defined the British Imperial Century.

André Paul Guillaume Gide, a French Nobel Prize laureate, writer and world traveler, said, "One doesn't discover new lands without consenting to lose sight, for a very long time, of the shore," (from The Counterfeiters, 1925) and from the latter part of the 18th century onwards many young Scotsmen did exactly that. They left their native shores, and the only home they had known, to seek knowledge and adventure elsewhere. This is the story of the creation of a small group of highly educated, self-sufficient, daring Scottish physicians and surgeons who led Great Britain in exploring the expanding empire.

1

Wars, Alliances and the Scottish Identity

> For so long as there shall but one hundred of us remain alive
> we will never give consent to subject ourselves to the dominion of the English.
> For it is not glory, it is not riches,
> neither is it honours, but it is liberty alone that we fight for,
> which no honest man will lose but with his life.

–The Declaration of Arbroath, 1320

Beginning with the Battle of Brunanburh in 937 CE, England and Scotland fought almost continuously for the next 800 years. At Brunanburh, although the forces of the King of England were opposed by the combined forces of the King of Dublin, the King of Scotland, and the King of Strathclyde,* the English won, and Michael Livingston states that this "was the moment when Englishness came of age." This initial conflict was the first step by both England and Scotland in creating their own fierce cultural identities, but the English victory did not curtail the ongoing attempts by Scotland to create an identity separate from her English neighbor. Livingston argues that the men who fought and died on that field forged a political map of the future that remains (in modernity), making the Battle of Brunanburh one of the most significant battles in the long history not just of England, but of the whole British Isles.[7] When considering the turbulent history of England and Scotland, their eventual merger into what would become the Kingdom of Great Britain seems implausible. The fact that Scotland would play such a disproportionate role in settling Britain's vast colonial empire, or that a group of Scottish doctors would come forward to lead the exploration on behalf of their combined nations, is even more difficult to understand. To appreciate how Scotland could move from being England's bitter enemy into the vanguard of leadership within the British Imperial Century requires a brief examination of the creation of a Scottish identity before and after the Acts of Union in 1707.

---

* One of the early medieval kingdoms of the Britons, located in what is now southern Scotland and part of Cumbria.

Creation of a national identity

Helen Dingwall, in *A History of Scottish Medicine: Themes and Influences*, describes the development of Scotland as a nation and the Scottish as a nationality as bringing together a collection of very separate, and locally controlled, social spheres. In order to forge an identity that was uniquely Scottish and separate from England, attempts first had to be made to unite these isolated spheres. But these ongoing attempts to create a unified nation were viewed as threats by the English, resulting in a series of battles that lasted from 1296 until 1357, which became known in Scotland as the First and Second Scottish Wars of Independence. At the end of the prolonged conflict, greatly assisted by Robert the Bruce's **(Figure 1.1)** victory at the Battle of Bannockburn in 1314, England was forced to acknowledge Scotland as an independent nation. Bruce ruled as Robert I until his untimely death in 1329; he was followed by his son David. The House of Bruce ruled from 1306 until 1371 and did much to establish the beginnings of a separate Scottish identity.[8] But then, as David died without heirs, the House of Bruce was followed by the House of Stuart, whose leaders ruled Scotland from 1371 until 1707. Following the Union of Crowns, the Stuarts continued to rule both Scotland and England until 1714. This relatively stable dynastic progression over more than three centuries made Scotland a more or less settled territory and continued strengthening the Scottish national character. This emergence of a uniquely Scottish identity is reflected in the characteristics of culture, language, and traditions which formed the basis for a legal structure, educational system, and religion quite separate from that of the Scots' southern neighbor. It was from these differences that the Scottish Enlightenment and the medical explorers would emerge.

England's acknowledgement of Scotland as a separate nation, and the relative stability of the Scottish monarchy, did not, however, stop the ongoing conflicts between these two neighbors, and the tides of battle between England and Scotland continued to ebb and flow. Sometimes, when a brilliant leader would emerge, the Scots tasted victory. More often the larger and usually better equipped English military, with a strong navy that could blockade ports and shell coastal communities, would beat the Scots once again into submission. It soon became apparent that English naval power would constantly keep the Scots at bay. What the Scots needed was a strong ally with a strong navy.

Figure 1.1 Robert the Bruce (Wikimedia Commons/PD)

The Auld Alliance

In many ways Scotland at the end of the 13th century was much more like Continental Europe than neighboring England, with whom the Scots shared a common border. While coastal and overland trade between Scotland and England was important, it was constantly being disrupted by downward spirals in cross-border relations between the two nations. As a result, Scotland's main source of trade was with northern Europe, specifically Denmark, France, and Holland, and it was natural that Scotland would look for a military ally among its already established diplomatic and trade partners. Nothing brings two nations together like a common enemy, and Scotland found France to be a most willing ally in providing ongoing military support.

The alliance created in 1295 between the Kingdoms of Scotland and France—still in place in the mid-16th century and by then known as the Auld*Alliance—was much more a continuation of ongoing European relationships than a new beginning.[9] The pact was first and foremost a diplomatic and military relationship. The terms of the treaty stipulated that if either country were to be attacked by England the other country would invade English territory. This agreement was especially valuable to France, in that whenever the English attempted war on the Continent, they had to keep a large army in reserve in order to repel a Scottish attack that would soon come charging across their northern border in support of the French.

Back then, France was among the few countries that could match England's powerful navy. This provided much-needed protection for coastal Scotland, and for generations Scotland relied on French warships to protect its coastline. The provision of military support was not, however, one-way. Lacking a strong navy but with a formidable fighting army, Scotland could send thousands of men eager to assist France in the fight against their shared enemy. At the Battle of Baugé in 1421, during the Hundred Years' War between the English and the French, it is estimated that 12,000 Scots fought for France. These fierce warriors were instrumental in helping the French defeat the English on that day at Baugé, and Scottish soldiers would serve in this role repeatedly and effectively during the course of the Auld Alliance.

The Wars of Independence further strengthened the Scottish identity and created a nation ruled by Scottish kings. These rulers bolstered Scotland's ties to Europe in general and especially to France. These connections, together with the long-standing and important trading links, created a Scottish identity that over time grew ever more continental and less English, with religion playing a major role in Scotland's alliances. But at that time Europe was rapidly changing. The Reformation had created a colossal rift across the continent. Large sections of Europe remained staunchly Catholic while other countries bought wholeheartedly into the Protestant movement. Although initially remaining within the Catholic circle of nations, Scotland was soon on the path toward the Protestant Reformation, which took hold there in the mid-16th century. The Reformation brought to Scotland the development of culture and ideas in many forms and in many areas; it would change the face of the nation permanently, not just within the confines of religious belief and practice.[10]

Scotland's adoption of Protestant ideology and its rejection of Catholicism spelled the end of the Auld Alliance, as it created a now Protestant Scotland in contrast to a continuing Catholic France. This significantly curtailed collaboration between the two countries. The

---

* Scots for 'old'

partnership between Scotland and France had lasted for almost 300 years, and although the Auld Alliance was never officially revoked, it is considered to have ended in 1560 with the signing of the Treaty of Edinburgh, involving Scotland, England, and France. Queen Elizabeth I, seizing an opportunity to diminish France's affiliation with Scotland, directed her commissioners to create the Anglo-Scottish Accord. With this document and with the support of the Scottish Lords of the Congregation (a group of Protestant Scottish nobles), the representatives of King Francis II of France agreed that the Auld Alliance would be replaced by the newly created accord.[11]

England moves from Tudors to Stuarts

The House of Stuart had ruled Scotland since 1371. The reigns of the various Stuart monarchs, although at times turbulent, maintained steady progress, and by the beginning of the 17th century Scotland had evolved from a collection of very separate but interacting regional spheres of influence into a consolidated national domain.[12] As mentioned earlier, this relatively stable dynastic progression made Scotland a more or less settled territory, and continued strengthening the Scottish national character. But a crucial event in terms of the formation of a Scottish identity transpired at the beginning of the 17th century. This event, contrary to similar events in history, resulted in the monarch of a small nation succeeding to the throne of its larger neighbor.

On March 24, 1603, England's Queen Elizabeth I died, leaving no heirs, and after 118 years the reign of the House of Tudor ended. It was considered that in order to avoid a religious civil war it was vital for the throne of England to be occupied by a Protestant, and King James VI of Scotland, (**Figure 1.2**) a Stuart cousin, was Elizabeth's closest Protestant relative. When she died, he also became James I, ruler of England and Ireland.[13] The act of placing Scotland and England under the same monarch was known as the Union of Crowns. James would likely have favored a complete union of the two kingdoms. But the Crown of England and the Crown of Scotland would remain both distinct and separate. The union between the two nations was personal and

Figure 1.2 King James I/VI
(Wikimedia Commons/PD)

dynastic: personal because James' relationship with Elizabeth I had allowed him to become King of both Scotland and England; dynastic in that upon his death both crowns would pass to his heirs.

Following the Union of Crowns the Stuarts ruled both Scotland and England until 1714. The only break in the Stuart dynasty was the nine years (1649–1658) when Oliver Cromwell took the title of Lord Protector of the briefly formed republic known as the Commonwealth of England. But Scotland and England remained separate nations, with separate parliaments, and with independent courts and laws that were ruled by the same king. This left Scotland an anomaly. England was politically more significant, geographically larger, more populous, wealthier, and militarily more powerful. In virtually every situation that arose, the Stuart rulers placed England's interests ahead of those of Scotland.[14] Scotland and England did not merge at this time. Whenever he could, James used the name Great Britain when referring to the two countries. But despite James' best efforts to create a new imperial throne his English subjects had no wish to merge their identity with that of Scotland, and regardless of the Union of Crowns, Scotland remained both culturally and economically more European than English.[15]

The turbulent 17th century

The entire 17th century saw Scotland and England ruled by a single monarch. This ended a portion of the conflict between the two nations, but the century could not be described as a period of calm. Following the death of James I/VI, Charles I was next in the Stuart line, and he assumed the throne in 1625. Although born in Scotland, Charles had become estranged from his northern kingdom, and he then attempted to impose his religious policies on the Church of Scotland, known to the Scots as the Kirk. Most in Scotland saw the changes as vehicles for introducing Anglicanism to Scotland and taking the governance of the Kirk out of the hands of its Scottish leaders. Support in Scotland for the royal government quickly collapsed, and unrest spread throughout the country. This led to what was known as the Bishops' War (1639–1640). The Scottish troops, many of whom were veterans of the Thirty Years' War, had far better training than their English counterparts, so the Scots successfully held off the English armies, and in August 1640 invaded England, where they met virtually no resistance until reaching Newcastle upon Tyne. There the Scots defeated the English at the Battle of Newburn and began an occupation of Northumberland and Durham.[16]

The huge ongoing cost of the war with Scotland, the defeat of the English army at Newburn, and the Scottish army's continuing occupation of much of northern England caused the relations between Charles and his English Parliament to decline further. With neither faction able to reach agreement or compromise, the Royalists, who supported Charles, and those supporting the English Parliament were soon engaged in the First and Second English Civil Wars, which lasted from 1642 to 1649.

Initially the Scots supported the English Parliament and its armies, directed by Oliver Cromwell. Large numbers of Scottish troops helped swell Cromwell's ranks, and the military balance tipped in favor of Cromwell's forces. There followed a series of defeats for the Royalists. Charles took refuge in Oxford, and Cromwell's forces laid siege to the city. In April

1646 Charles escaped from Oxford and surrendered to a largely Scottish force who, after nine months of negotiations, turned him over to Cromwell. Almost at once the Scots began to have doubts about the role they had played in the king's capture. Following the Bishops' War, Charles had conceded to almost all the Scottish demands. This had divided the country over the alliance with Cromwell versus support of the Crown. These concerns proved to be well founded. After the king was turned over to Cromwell he was held captive for two years. During that time, he was twice brought before the High Court of Justice, and was sentenced to death. On January 30, 1649, the king was executed. Word of the execution was received in Scotland with dismay. Even those most strongly opposed to Charles and his policies felt shame and shock at the role that Scotland had played in removing the king.[17]

In 1649, following the king's execution, Cromwell established the Commonwealth of England as a replacement to the monarchy. The execution of the king and the founding of the Commonwealth fundamentally reshaped the political landscape in Scotland and prompted vast and significant concerns regarding continued allegiance to Cromwell. A growing number of Scots felt they had no option but to switch their support to the Royalists supporting Charles I's son and heir, Prince Charles. Tensions continued to increase, and ultimately led to a royal counter-revolution.[18] This put Scotland in opposition to both Parliament and Cromwell. Prince Charles was in exile, but returned to Scotland in June 1650, attempting to raise an army among his Scottish supporters. The war dragged on for a year until the Royalists were finally defeated at Worcester in 1651 and young Charles was forced to flee again to France.[19]

Cromwell continued to rule the Commonwealth until his death in 1658. At that point leadership passed to his son, Richard. However, Richard did not have the support of Parliament or the army and Prince Charles, having spent nine years in exile, was invited to return to Britain; in 1660 he became King Charles II. The new king was welcomed back with great public acclaim and ruled until his death in 1685. Charles received an especially warm welcome from the Scots, and by the end of the 17th century Scotland found itself moving toward an even tighter connection with England.

Stuarts to Hanoverians

The first two Stuart kings of England, James I and Charles I, were Protestant. Charles II, while publicly declaring that he was Protestant, leaned towards Catholicism in private.* Charles II's brother, James, the next in line for the throne should Charles leave no heir, was initially Protestant. For reasons which are not entirely clear, while Charles II was still the reigning monarch, both James and his wife converted to the Roman Catholic faith. In 1671 James' wife died and, having already made one unpopular decision—his conversion—within Protestant Britain, he made another. He began scouring the courts of Europe in search of a Catholic princess to be his second wife. In Italy he found Mary of Modena, and they married. Had Mary been Protestant she most likely would have been welcomed

---

* Charles had been raised in Catholic France, and had married a Catholic, and then in 1670 he signed a secret treaty with Louis XIV, promising to convert to Catholicism. Finally on his deathbed he openly converted to Catholicism.

as the future queen. She was beautiful and possessed great charm, but as a Catholic she immediately aroused suspicion. But gradually, as the reign of Charles II continued, James and Mary began to win over public opinion. James was a conservative, a traditionalist, and a disciplinarian. His wife was beautiful and brought glamour to the court, and neither of them flaunted their religion. Charles II had seen to it that the daughters from James' first marriage, Mary and Anne, had been raised Protestant. Now, since James was nearing 50, and Mary of Modena's eight pregnancies had resulted in a series of miscarriages and stillbirths, it was unlikely that an heir from this, his second marriage, would be forthcoming.[20]

In 1685, upon Charles II's death, James II became King of England and Scotland, continuing the reign of the House of Stuart. For the first time in nearly 120 years, Scotland and England now had a Roman Catholic monarch. And at the time of his accession, James seemed destined for a successful reign. His religion did not appear to be a critical issue, as his daughter Mary, from his first marriage, was next in line for the throne. Now an adult, Mary was an Anglican and married to William of Orange, her cousin, who was Stadtholder (President) of the Dutch United Provinces and also a Protestant.

However, the birth of a son to James in June 1688 effectively excluded Mary from the line of succession. This greatly troubled the political leaders of the time, as it created the potential of a second Catholic ruler for both England and Scotland and could lead to an attempt to reestablish Catholicism throughout both domains.[21] Prominent powerful Englishmen felt they could not allow this to happen. A delegation traveled to Holland, and William was given to understand that if he advanced on England with an army he would be "met with considerable support." [22] William was half-Stuart, married to a Stuart, a Protestant, and he felt he was entitled to the crown. He invaded England and advanced with his army toward London. But before a battle could ensue, James, and his wife and son, escaped into exile. In 1689, within six weeks of James' arrival in France, William and Mary jointly assumed the vacated throne. This became known as the Glorious Revolution.[23]

But then in 1694 Queen Mary died of smallpox; she was only 32 years old. William continued to reign alone into the beginning of the 18th century. As William and Mary had no children, Anne, also an Anglican and a daughter from James' first marriage, was next in the line of secession. To this point, Anne had borne no less than 17 children. However, most of them had died in infancy or at birth. Only her son William had survived beyond his early years, but in 1700 he too died, at age 11.

Many of the Stuart cousins were Catholic, and the leading politicians again feared that continuation of the House of Stuart would result in a succession of Roman Catholic kings and queens. Shortly after the death of Anne's son, King William met with his cousin Sophia of Hanover, suggesting that she be inserted into England's line of succession, and she agreed. The decision passed over 56 Catholic Stuart relatives with stronger hereditary claims.

Sophia was German, and not a member of the House of Stuart—but she was Protestant, as was her son. William's decision to change the line of succession and end the reign of the House of Stuart was implemented in June 1701 by the English Parliament in the Act of Settlement. This document stated that in the event of no legitimate issue from William III or Anne, the crowns of England, Ireland, and Wales would pass to "the most excellent princess

Figure 1.3 Queen Anne
(Wikimedia Commons/PD)

Sophia, electress and duchess-dowager of Hanover" and "the heirs of her body, being Protestant." [24] It is notable that the Act of Settlement did not mention Scotland.

Sophia of Hanover and her heirs were now designated to follow Anne onto the English throne. But this only impacted England, Ireland, and Wales, all of whom adhered to the English Parliament. Scotland was still operating with its own independent parliament and could vote whether or not to accept the new Hanoverian line. The Scots could instead invite the titular Prince of Wales, son of the exiled James, to the throne, and if that happened the two kingdoms would then again be separated after being under the same ruler for a century.[25] If England attempted to impose this new line of succession on Scotland through force, war could once more ensue. But as England was already at war with France, and its armies and finances were stretched thin, a war on its northern border could be disastrous for it.[26]

In 1702, upon William's death, Anne **(Figure 1.3)** became queen. She knew that the only way to guarantee a smooth, and Protestant, transition upon the completion of her reign was to act quickly to remove one of the final vestiges of Scottish independence. The House of Hanover could not assume both crowns without separate approval by the Scottish Parliament. To Anne, the simplest solution was that it had to be dissolved. From the beginning of her reign, through skilled political maneuvering, and more than a touch of bribery, Queen Anne began putting the final step into place that would ensure Protestant lineage. That move would be a treaty to become known as the First Act of Union, which would incorporate the Kingdom of Scotland, along with Wales and England, into the Kingdom of Great Britain under a single parliament. This new parliament would rule from London.

From October 1706 until January 1707, the Scottish Parliament voted, article by article, to ratify the treaty as proposed by Queen Anne. A large portion of Scotland was opposed to the House of Stuart being replaced by the House of Hanover; it was one thing to be ruled by a Scottish king who lived in London but quite another to be ruled by a German king who didn't even speak English. Also, a large percentage of the Scottish population believed that a handful of landed and affluent Scots were pressing for the merger purely for their own financial advantage, and that these men were willing to sacrifice Scotland for personal gain. As Scotland moved toward unification with England dozens of petitions against the merger rained down on the Scottish Parliament. Riots erupted in Edinburgh and Glasgow, and angry demonstrators burned copies of the treaty in Dumfries and Stirling.[27] But despite the protests, on January 16, 1707, the Second Act of Union, merging the two Kingdoms of Scotland and England, was ratified by the Scottish Parliament.

Linda Colley states that the kingdom that emerged in 1707 was consolidated around what Britons had in common rather than what divided them. At the heart of this self-definition was Protestantism, "creating in Great Britain an island of Protestantism circled by a seething Catholic sea." [28] Now under a single ruler and a single parliament, these separate nations became in the words of the Union Treaty, "United into One Kingdom by the Name of Great Britain."

Queen Anne, the last of the Stuart rulers, died in 1714. Upon her death George I, the eldest son of Sophia of Hanover, became ruler of the Kingdom of Great Britain. The reign of the Stuarts had ended, and the Hanoverian reign had begun.

Conquest, absorption or merger?

At the beginning of the 18th century, England had five times the population of Scotland and nearly 36 times its wealth.[29] Scotland's decision to accept the English Act of Union and write its own complementary Act was based on economics; the Scots ultimately approved the pact in an attempt to address their vast and varied financial problems. Without the union, the potential for the Scots to boost their economy was not promising. From the 14th century onward wars, plagues, drought, famine, and other scourges had affected most parts of the country. These severe aspects caused many to leave the country, resulting in losses in the general population. But these factors also limited the products that the Scots were able to trade. In 1700 urbanization and its accompanying boom in manufacturing, which was spreading across the rest of Europe, was virtually nonexistent within Scotland. Edinburgh (including its port of Leith), with a population of approximately 50,000, was the only city of any size. Scottish exports were primarily products of the land such as wool, hides, and fish. In addition to a slump in trade, food shortages had created widespread famine. These factors, along with rampant inflation, created a desperate situation in Scotland, so it was believed by many that a complete merger with England could effectively address the Scottish plight.[30]

For the English, the merger would provide protection for their northern border. Although Scotland and England had essentially been united for over a century, the border hostilities had continued. Scotland had periodically continued to align itself militarily with France. England sought the merger to create a more defensible single island and to put an end to the 17th century's series of ongoing battles with the Scots. Perhaps more importantly it also protected the continuation of the Protestant line of succession. Never again would a Catholic monarch ascend to the combined thrones. If this appeared to be happening the Kingdom of Great Britain, now operating under a combined parliament, could move to act.

The union of Scotland and England was debated during the merger of 1707, and for long after it was implemented. The factors leading to the Scottish acceptance of the English terms, the methodology employed, and the ultimate benefits and losses to the people of Scotland were all points of contention. What was the least contentious was that national commerce in Scotland had stagnated and perhaps this was the most compelling motivation for the merger. The bulk of the articles within the treaty refer to economic issues that would ultimately prove favorable to the Scots.[31] The Scottish economy benefited in a number of ways from the Acts of Union. Throughout the period leading to the signing, the economy had been precarious, with rampant inflation in the 15th and 16th centuries, and limited recovery during the 17th century punctuated by periods of plague and harvest failures, together with the politics of restrictive trade barriers. But this gave way to a more stable economic environment.[32]

A major benefit of the merger was Scotland's unrestricted access to English markets. Banking, too, developed during this period. While the Bank of Scotland had been founded shortly before the union, in 1727 the Royal Bank of Scotland followed, with local banks beginning to be established in Glasgow and Ayr. This made capital available for business and trade, and also led to more money being made available for improvement in the infrastructure, leading to better roads, canals, and harbors.[33] The new access to travel and developing commerce began to unite the many isolated pockets within the country, and this activity further promoted the rapid development of the Scottish identity.

The improvement in transportation can be seen clearly in Rosemary Goring's *Scotland: The Autobiography*. In 1763:

> There were two stagecoaches which went to Leith (a port area near Edinburgh) every hour from eight in the morning till eight at night. There were no other stagecoaches in Scotland, except one, which set out once a month for London, and it was from twelve to sixteen days upon the journey.

Then in 1783:

> There were six stagecoaches to Leith every half hour. There were stagecoaches, flies (a one-horse hackney carriage), and diligences (a public coach) to every considerable town in Scotland, and to many of them two, three, four and five per week. To London there were sixty stagecoaches monthly and the capital was reached in four days.[34]

Peter and Fiona Somerset Fry add:

> Reflecting today, the merging of the two countries has proven to be a beneficial economic and political advantage for Scotland as well as England, although it was a union that few people in Scotland wanted at the time.[35]

The early years after the Acts of Union proved difficult, but economic expansion soon allowed Scotland to close the financial gap with England. The production of textiles, already a significant Scottish industry, expanded dramatically. Although trade with England was beneficial, the major change was in international trade and the expansion of the markets within the East Indies, the West Indies, and the North American colonies, where a highly successful tobacco trade emerged.[36] By the middle of the 18th century the union of the two countries that had produced Scotland's economic growth had also resulted in a huge increase in the population of Scotland's major cities. By 1851 one-third of the country's population was classified as urban. Edinburgh's population grew from 52,000 to 83,000. Dundee's population doubled, and Aberdeen grew by 80 percent.

But the growth in Glasgow was the most remarkable. The strong economic ties with Europe would remain, but the shift in focus to the Americas helped to ensure that Glasgow and the west of Scotland were the prime locations of the economic boom. At the time of the merger, in 1707, Glasgow had a population of approximately 13,000. By the end of the 18th century, it had grown to over 54,000 people. In the 1820 census it had a population of 147,000; in 1901 it had grown to 762,000. Glasgow—now nearly *60 times* bigger than in 1707—had become the "second greatest city in Britain" and had overtaken Bristol as the major shipping port.[37] Between 1740 and 1820 these rapidly forming urban centers would become the focal points for the Scottish Enlightenment.

The primary catalyst for the peaceful union between Scotland and England was the succession to the throne. Unlike Ireland and Wales, which also came under England's control, Scotland was not annexed, it was not conquered, and it was not absorbed. England and Scotland were merged.[38] This political agreement, as opposed to annexation or military conquest, allowed Scotland to enter the accord without a loss of identity. It maintained its strong sense of Scottishness. According to Karin Bowie, Scotland brought into the union "a sense of itself as a national political community. This ensured that a contemporary 'sense of nation' could be sustained and developed" following the merger.[39]

The stage was set, and the Scottish Enlightenment would take place in the context and atmosphere of an urban society that was rapidly growing, improving, and developing, with a sense of who it was and what it was to become. Each side gained from the merger. How this union had been forged, and whether it had been done against the will of the people, became a moot point. The closer and more cooperative contact between England and Scotland, which had originated with the Union of Crowns, was followed by the Acts of Union, which merged the two nations' parliaments. From that point forward, the fate of Scotland was tied to that of England. Scotland would not only be part of a kingdom comprised of multiple nations but would also emerge as a significant partner in pursuit of Great Britain's empire building. Scotland would offer countless opportunities for educated, ambitious, and mobile young Scots in careers never before imagined. The task of empire building would emerge from the study of medicine.

2

The Athens of the North

Today it is from Scotland that we get rules of taste in all the arts,
from epic poetry to gardening. . . We look to Scotland
for all our ideas of civilization."

–Voltaire

The Scottish Enlightenment swept the Western World with a whirlwind of thoughts, ideas, and theorics, and with an intellectual vigor that would place Scotland at the forefront of scholarly enterprises for the 80 years leading up to the British Imperial Century. As the Enlightenment unfolded there were visible outcomes that were recognized throughout Europe. But there were also outcomes that were not to be known until years after the movement had ceased to be. As the Scottish Enlightenment was ending, the British Imperial Century was beginning. The Enlightenment had established the foundation for the evolution of a singular group of explorers who would emerge from obscurity in Scotland and travel the globe while collecting, documenting, charting, and sharing their investigations of the natural world. They would explore like no one before —with knowledge, vision, strength, and humanity. Without the Scottish Enlightenment there would have been no medical explorers.

Of institutions and ideas

Jean-Jacques Rousseau was holding court in a candlelit French salon. It was 1761, and he had gathered a few friends to share some thoughts he had assembled for his forthcoming book, *Emile, or On Education.*

> To live is not to breathe but to act. It is to make use of our organs, our senses, our faculties, of all the parts of ourselves which give us the sentiment of our existence. The man who has lived the most is not he who has counted the most years but he who has most felt life.

Emile would be published the following year and become an immediate commercial success. It would be considered one of the seminal works to come out of the Age of Enlightenment. It would also infuriate Protestants and Catholics alike. It outraged the French Parliament to the point that it issued an arrest order against Rousseau. His contemporary and philosophical rival Voltaire described the section titled 'Profession of Faith of the Savoyard Vicar' as "forty pages against Christianity … He says as many hurtful things against the philosophers as against Jesus Christ, but the philosophers will be more indulgent than the priests." [40] The book would be banned in Paris and Geneva, and often publicly burned upon its introduction into a new city. But during the French Revolution, some 30 years later, *Emile* would be the inspiration for what would become a new national system for French education. Rousseau's observation that "without having a broad range of thoughts and ideas about how children and young adults develop, it is easy to get locked into a rigid and inflexible system that ignores the importance of creativity and freedom" would transform French education.[41] His philosophy of education would also be a major influence in the revision of Scotland's university system during the Scottish Enlightenment.

Historians traditionally date the Age of Enlightenment as between 1715 and 1789. It was an intellectual and philosophical movement that opened minds to new ways of examining the world. This movement was centered on the European continent, most specifically in France, and was a time that saw the emergence of economics, history, sociology, and psychology as newly independent disciplines. Great thinkers emerged, spreading innovation and creativity. When one mentions the Age of Enlightenment to most people now, this conjures up images of Rousseau and his fellow countryman, Voltaire. But back then, in the 18th century, it was the German philosopher Immanuel Kant, the Italian jurist Cesare Beccaria, and the Dutch theologian Hugo Grotius who seemed to dominate every discussion of culture and learning. The philosophic and scientific activities that challenged the traditional doctrines and dogmas were not limited to Paris, nor even to the Continent as a whole. The Age of Enlightenment would forever impact human thought, but it would also usher in a Scottish age which produced its own era of insight and productivity.

While the activities of the Age of Enlightenment began on the European Continent, its work overlapped the Scottish Enlightenment, which is generally dated between 1740 and 1820. The European Enlightenment was an era of reason and debate throughout Europe, and Scotland played a significant role in its formation. Although the Scottish Enlightenment is often overlooked in comparison to its continental counterpart, to discount its role is a mistake. According to Arthur Herman in *How the Scots Invented the Modern World*, "the Scottish Enlightenment may have been less glamorous than its French counterpart but in many ways was more robust and original." [42] In both movements, new and exceptional thought and cultural expression laid a critical intellectual foundation for the modern world and this was a time in which Scottish scholarly life flourished as never before. Voltaire described Edinburgh as an intellectual center that rivaled Paris, London, and Vienna. Thomas Jefferson, the great American intellect, said that in science, "no place in the world can compete with Edinburgh." [43]

Like the Continental Age of Enlightenment, which contributed to and often redefined Europe's understanding of philosophy, economics, architecture, art, and literature, the

Scottish Enlightenment had a substantial impact in these fields. In Scotland David Hume (philosophy), Adam Smith (economics), Robert Burns (literature), Adam Ferguson (sociology), Colin Maclaurin (mathematics), Joseph Black (medicine and chemistry) and James Hutton (geology) were recognized around the world as leaders in their respective disciplines. But in addition, the Scottish Enlightenment also provided great vision in the fields of natural and physical sciences, and to the study of medicine.

Christopher Berry defines the Scottish movement in *The Idea of Commercial Society in the Scottish Enlightenment* with the following. "The Scottish Enlightenment is both a set of institutions and a set of ideas. As such they represent two differing facets of a complex whole." [44] The institutions to which Berry refers are the three central agencies that played conspicuous roles in Scotland's history: the Church of Scotland (also known by its Scots name, the Kirk), the Scottish legal system and the Scottish universities. The contributors to the European Age of Enlightenment were participants in the French salons, but in the Scottish movement discussions were found in the halls of Scottish universities, in Scottish legal and ecclesiastical circles and within the membership of societies and clubs. [45] The seats of governance for each of these institutions were in the cities, and the Scottish Enlightenment was primarily an urban movement in that it was facilitated by the forms of social and intellectual expression that city living encouraged. The Scottish Enlightenment was felt most strongly in the urban areas within the lowland belt bounded by Glasgow in the west and Edinburgh in the east. It also included the city of Aberdeen. Because of the location of the country's major cities, the Scottish Enlightenment did not impact all areas of Scotland equally, and the Lowlands were far more involved than the Highlands.

The last, and perhaps arguably the most important factor producing the Scottish Enlightenment was the Acts of Union between England and Scotland in 1707 which created a single capital and single parliament for the two countries. The Scottish parliament ceased to exist and those who had supported the earlier governmental efforts of Scotland —that is, the prominent politicians and office holders —moved to the new center of power in London. Nicholas Phillipson speculates this had a "traumatic effect" on the population in that the elite of Scottish society had come primarily from those controlling the government. The Scottish nation, now "bereft of political institutions and dissatisfied with the remnants of an ancient Scottish culture," needed to define a new societal order. [46] Scottish law and the Church of Scotland remained, however, untouched by the dissolving of the parliament, as did the great universities located in St. Andrews, Glasgow, Aberdeen, and Edinburgh. The legal profession, the leadership of the church, and those within the university academic communities would unite to fill the void created by the parliamentary exodus. According to Phillipson and other academics, the Scottish Enlightenment would grow from these unique entities combining to develop a distinctive Scottish national culture, almost in protest against the loss of nationhood entailed by the Acts of Union.

The lawyers, church leaders, and university professors became known as the intelligentsia and formed the nucleus for the Scottish Enlightenment just as the cities, which would create their platform, came into existence. For much of its history, Scotland had little need for large settlements. Throughout most of the country its inhabitants experienced a rural life that involved small, tilled acreages and shared pasture lands located around small villages. At the time of the Acts of Union, probably no more than 10 percent of the population of Scotland lived in towns of any size.[47] Urban development began during the 18th century when the Industrial Revolution created a demand for labor. The process of urbanization gathered pace, and by the start of the Scottish Enlightenment the diverse triad of institutions were thriving in their newly expanded urban settings. The four university cities all experienced increases in population. The concentration of Scottish lawyers, clerical leaders, and university professors in each of these city centers soon attracted other intellectuals, including physicians, scientists, and architects, creating a powerhouse of Scottish thought that dominated urban society in Scotland. It was the collaboration between these intellectuals that created Berry's "set of ideas" that then ignited the Scottish Enlightenment.

Although the Enlightenment came to an end in Scotland in about 1820, the trend toward urbanization did not. Well before the Great War, Scotland had emerged as one of the world's most highly urbanized societies. By 1850, in the midst of the Imperial Century, over 20 percent of Scots lived in the ancient and prestigious royal burghs* of Glasgow, Edinburgh, Dundee, or Aberdeen, and by the end of the Imperial Century over one in three Scots was an urban dweller.[48] Soon, 50 percent of the population was living in towns of 20,000 or more, and only England exceeded Scotland as the most urbanized nation in the world.[49] Between 1770 and 1800 the population of Edinburgh doubled to 80,000, and through the 19th century it continued its dramatic rise. By 1911 nearly 60 percent of Scots lived in settlements with a population of 5,000 or greater.[50]

Although Glasgow experienced one of the fastest-growing populations in the United Kingdom and became the largest city in Scotland, Edinburgh quickly emerged as the greatest among the urban university centers. From the middle of the 17th century the city became the professional and intellectual capital of Scotland. Edinburgh was favorably placed as the home of its highly regarded university. It was also the seat of the Scottish legal system, and the venue of the General Assembly of the Kirk.[51] Boasting over 200 representatives of the legal profession alone, by the end of the 17th century Edinburgh's professionals greatly outnumbered its merchants. The legal professionals would join forces with the leaders of the Kirk and the university academics, and these institutions and their set of ideas would enable the Scottish Enlightenment to propel the nation to greatness. At the turn of the 18th century great things were poised to happen in Scotland, and they would happen most often in Edinburgh.

The Kirk

It may be said that the progress of any nation is affected deeply by the religious persuasion of its inhabitants, and Scotland is a prime example. The history of religion among the people of Scotland is one of evolution through the first two millennia CE. According to Roman accounts, upon their arrival in approximately 50 BCE Britain, Ireland, and Scotland practiced Celtic

---

* A burgh was a town incorporated by royal charter, which enjoyed a degree of self-government

paganism. This practice was described as "a basic religious homogeneity" led by "magico-religious specialists" known as the druids. Within 100 years of the Romans' arrival, Celtic religious practices began to display elements of Romanization, resulting in a hybrid Celtic faith with its own set of deities.[52] The Romans' original plan had been to conquer and occupy all of Scotland, but the need to protect the homeland made it necessary for Rome to withdraw its legions from throughout southern Scotland and northern England before the conquest could be completed. By 430 CE the last remaining Roman outposts in Scotland had been abandoned and Scottish society had begun to fragment into a maze of warring minor states. But there were still memories of a Christian Roman past.[53]

By the 6th century the ancient Celtic worship had been completely overtaken by the Roman Catholic Church. The influence of Catholicism became nearly all-consuming, with the "politics of religion and the religion of politics" almost indistinguishable through the progression of a series of Roman Catholic rulers.[54] By 800 CE, while some elements of earlier society survived, much had changed; tribal leaders were starting to become dynastic kings of regions rather than heads of related tribes, their power and wealth was increasing, and new forms of government were being introduced.[55] This system of regional leaders continued until the middle of the 9th century, when Kenneth I became the first king of a united Scotland. Until the Protestant Reformation of the mid-16th century, control of the nation continued under a series of Roman Catholic rulers. Although the Reformation brought change to Scotland, the change did not occur quickly or easily.

Scottish domestic and international policy was affected by the Reformation, especially in 1534, when the English King Henry VIII broke from the Roman Catholic Church and formed the Church of England. However, the Scottish King James V did not follow suit. Rather, he aggressively attempted to stamp out Protestantism in Scotland, and a number of outspoken Scottish Protestants were convicted of heresy and burned at the stake. The English ambassador to Scotland tried to encourage James to close the monasteries, but James instead lobbied Pope Clement VII to allow him to tax the monastic incomes. The Pope, apparently seeing this request as preferable to Scotland breaking from the Church as England had done, readily agreed. But this adherence to the Roman Catholic Church was not to survive; by 1603 and the Union of Crowns, Protestantism had replaced Catholicism among the reigning Scottish monarchs. Then the introduction of the House of Hanover into the English line of succession at the beginning of the 18th century, followed by the Acts of Union in 1707, guaranteed that Scotland would be part of the Kingdom of Great Britain, and that from that point forward the monarch of this kingdom would be Protestant. The shift in alliance and allegiance from Catholic to Protestant would in the long-term help mold the Scottish nation and character, making the Church of Scotland one of the three guiding factors within the Scottish Enlightenment.

The Acts of Union, protecting the Church of Scotland, promised that the merger would not alter "Worship, Discipline and Government of the Church of this Kingdom." [56] Henry VIII's Church of England had become the "established church of the country," and it certainly was important to vast aspects of law and society in England. In Scotland, however, to an even greater extent, religion, led by the Church of Scotland, would be seen to dominate every utterance and act of the Scottish people. The basic unit of the Church of Scotland

Figure 2.1 General Assembly of the Kirk (Wikimedia Commons/PD)

was, within each district, the local Kirk session, led by a presbytery, an administrative body representing all the congregations within that locality. A group of presbyteries formed a synod, an ecclesiastical court above the presbyteries and subject to the General Assembly. The latter was the supreme body, and most of its members came from the presbyteries. However, governmental representatives from the royal burghs and representatives selected from within the leadership of the Scottish universities were also included within the General Assembly. **(Figure 2.1)** Life in Scotland was centered in terms of the Church and its role in the people's lives. Its teachings spilled over into politics, domestic life, and social activities. Through the Church, the Scots saw it as their destiny to improve every aspect of society: public and private conduct, trade and industry, agriculture, and education.[57] The way the Church of Scotland would accomplish this was to become one-third of the powerful and influential trichotomy leading the Scottish Enlightenment.

Although in the early days of the Enlightenment the Church of Scotland was fully established, it had begun to fray at the edges; the aging links between religion and politics were beginning to pull the church apart. Both the leadership and the general membership of the Church of Scotland were divided into two quite different factions, the Moderates and the Evangelicals; the primary difference between the two would be expressed in a battle over how the Kirk would respond to the societal and intellectual shifts in Scotland brought on by the Enlightenment.

The Evangelicals emphasized the Bible, the doctrine of atonement, conversion of others to the faith, and the need to spread the gospel. They looked outward, as missionaries working to convert the greater world. The Evangelicals, in addition, saw the Church's collaboration with the lawyers and academics, whose members were generally drawn from the upper strata of society, simply as an attempt by the wealthy non-theologians to exert increasing control over the Church.[58]

The Moderates were, however, more concerned with the general public of Scotland. They wished the Church to continue to be a central and stabilizing influence in a society that was undergoing massive economic and social changes.

The Moderates believed that for the Church to relate effectively to the changing society of the mid-18th century, they would need to ally with fields of thought outside of theology, such as the law and in academia. The Church, led by the Moderates, realized that these powerful non-ecclesiastic forces were strongly influencing society and saw it as vital for the Church to work in concert with the lawyers and professors rather than opposing them. From about 1750 through the remainder of the Enlightenment period, the General Assembly of the Church of Scotland was dominated by men representing the Moderate Party of the Church, who fought successfully to ensure that they could work effectively within the changing society. The Moderates' movement in support of the Enlightenment, to the exclusion of the Evangelicals, was known as the "Improvement." With this, the Church of Scotland and the Enlightenment were soon completely interwoven and inseparable.

The Church continued to work to improve Scottish society by encouraging advancements in literacy, science, and the arts. Influential Church leaders such as William Robertson became significant figures in the Scottish Enlightenment by firmly embedding the Moderate wing of the Church of Scotland into the movement. Ordained as a minister in the Church of Scotland, he soon became a member of the Church's General Assembly. He not only supported the vision of the Enlightenment but contributed to its body of literature. Educated at the University of Edinburgh, he is regarded as one of the most important historians of the 18th century; his publications established his reputation as a historian, and in 1762 he was appointed Principal of the University of Edinburgh. In 1763 he was named Moderator of the General Assembly. He would go on to be named Historiographer Royal for Scotland and become one of the founding members of the Royal Society of Edinburgh when it received its Royal Charter in 1783. His position as moderator, as well as the influence provided through his appointment as principal, allowed him to tie the views of the Moderates within the Church to the educational practices of the university.

The personal involvement of men such as William Robertson, combined with the General Assembly's conducive location in Edinburgh, ensured that the Moderates' control of Church affairs would support the intellectual movement then consuming the Scottish people, and especially the Edinburgh intellectual elite.[59] But following the Enlightenment the differences between the Evangelicals and Moderates widened. Seeing little hope in changing the system from within, and after much thought and debate, the Evangelicals decided to break away from the Church of Scotland. In 1843, after years of growing discontent, over 40 ministers —about 40 percent of those within the church at the time—decided to form the Free Church of Scotland. This was known as the "Great Disruption." It was not until 1900, at almost the end of the British Imperial Century, that the first major attempt was made to heal the rift. Meanwhile this separation had left the Church of Scotland's Moderate philosophy as the guiding principle for the Enlightenment and for most of the Imperial Century that followed.[60]

The law

The most dramatic impact of the Acts of Union was that the newly formed Kingdom of Great Britain "was to be represented by one and the same parliament." The treaty gave the Scots little direct power (only 8 percent in both the Lords and the Commons). But, perhaps more importantly, it did allow the Scots to retain their legal system, which was much more closely linked to European/Roman law than to English common law. Scottish law had far stronger connections to the rest of Europe than to England. In fact, until the creation of law chairs within the Scottish universities, those practicing law in Scotland had generally received their training in the Dutch universities of Leiden and Utrecht.[61] Those returning after completing their legal training abroad had been broadly exposed to the European Enlightenment's thoughts and practices through their education and travels. This created a large cohort of well-to-do, well-educated individuals poised to be included among the powerful and influential leadership of the Scottish Enlightenment.

The practice of law in Scotland was the domain of the wealthy. Earning a degree at a European university, with the cost of travel, housing and maintenance, and university fees, was extremely expensive. To become a lawyer was to attain the most influential and financially rewarding profession in Scotland, and the study of law was reserved for aspirants from socially established, even aristocratic, families.[62] The reasons for this were financial, but also social. The legal practitioners were crucial to agricultural improvement, industrial and commercial development, and politics—indeed, most spheres of Scottish social life in that period. The lawyers and the legal system were the center of an organization that touched activities conducted by Scotland's prosperous upper and middle classes. Only a career at the bar gave one such social, political, and financial access and advantages. James Boswell, later to become Samuel Johnson's biographer, wrote that "the high society of Edinburgh consisted of lawyers, and they alone set the social tone." [63]

The influence that the lawyers wielded over Scottish society only slightly lessened with the creation of law programs within the Scottish universities. After the study of law became available in Scotland, exposure to European thought continued to be advanced through the university professors who, for the majority, had been trained in Holland and France. Studying law in Scotland still required wealth; however, with universal male education, and universities accepting students regardless of social or economic status, the students studying law were no longer exclusively the offspring of the upper class but included the sons of the professional and merchant classes as well.

The societal position of the legal profession during the Scottish Enlightenment continued to grow in both prestige and wealth. With the Acts of Union, the landed gentry who had comprised the Scottish Parliament left Edinburgh for London, taking with them the prestige associated with their office. The legal profession was not affected by either the Union of Crowns or the Acts of Union, and with the parliamentary exodus to London the lawyers became the most prominent and esteemed occupation that the country knew. Lawyers set the nature and pace of society, and the revenue they received from the land and commercial developments provided them with the trappings crucial to social acceptance. Many of the Enlightenment's most brilliant thinkers were lawyers, including John Millar, Henry Home, and James Burnett. In addition to the General Assembly of the Church of Scotland, the location of the Supreme Courts of Scotland was also Edinburgh. **(Figure 2.2)** This ensured that these men of intellect, leisure, and wealth would place their profession and the city at the heart of the Enlightenment. Their collaboration with the Church and the universities would also bear out the validity of Cockburn's quote that "the law was a profession congenial to the practice of literature and philosophy." [64]

The universities

The primary contribution of the Church to the Scottish Enlightenment was to set a societal moral standard well founded on centuries of prudent custom centered within the Church of Scotland. The inclusion of the legal profession within the leadership of the Enlightenment ensured the commitment of men who represented the apex of this society through their prestige and wealth. As important as the law and the Church were to the Scottish Enlightenment, the principal thinkers had their institutional and intellectual base within the universities, and the roots of the Scottish Enlightenment lay deep within the nation's educational values. Education had for a long time been considered essential by the Scots, and despite the decades of poverty there remained in Scotland a tradition of scholarship and a hunger and respect for learning. By the late 17th century Scotland had

Figure 2.2 Parliament House Edinburgh (Wikimedia Commons/PD)

already developed a mandatory education system linked to the Church that established a school in every parish. Literacy and numeracy were much more widespread in Scotland than in neighboring England or in Germany and France. Well ahead of the English, the Scots had a mandatory education system that educated all males regardless of social class or financial status.[65] As a result most Scottish men were remarkably well informed on aspects of religion, current affairs, politics, and history, and the level of general literacy was extremely high. In 1796 the importance of an education to even the lowest classes of society was noted: "even day-labourers give their children a good education … an important advantage which the Scots as a nation enjoy over the natives of other countries."[66]

In the Lowlands there was a complete network of schools, though in many areas of the Highlands education was still lacking. Primary schools were set up in every presbytery, with grammar schools located in the larger towns. Heritors (substantial landowners) in each parish were expected to support the primary schools, and by the early 19th century these schools were generally open to girls as well as boys. As the Industrial Revolution rapidly increased the populations of the major Lowland cities, the public grammar schools found it more and more difficult to cope with the increased numbers of students. By 1818 nearly two-thirds of the school-aged children were gaining their education in private "adventure schools." These institutions were academically equal to the grammar schools,[*] and their control rested not with the Church but with the pupils' parents, which appealed to many families.[67]

Both the grammar schools and the private adventure schools in the larger towns provided a true college preparatory education. These programs ranged from four to seven years in length. Boys would enter at age nine or ten, having mastered their basics in primary schools. The secondary schools were not simply for the children of the rich, but were open to all boys, and a concerted effort was made to discover those with the most ability, regardless of their family's station or wealth. The fact that sons of poor farmers as well as the sons of the wealthy were literate meant that this universal education prepared capable boys of all social strata for a university education. For boys with talent but no resources, bursaries[†] provided by the universities were widely available.[68] The high regard for education, and the promotion of education for all, were central elements of the Scottish Enlightenment and transformed an already strong educational structure into the best system in Europe.[69] J. Currie, an early biographer of Robert Burns, recorded his observations regarding education among even the poorest Scots:

> A slight acquaintance with the peasantry of Scotland would service to convince the unprejudiced observer that they possess a degree of intelligence not generally found among the same class of men in other countries of Europe. In the very humblest of the Scottish peasants everyone can read, and most persons are more or less skilled in writing and arithmetic; and under the disguise of their uncouth appearance, and of the peculiar manners and dialect, a stranger will discover that they possess a curiosity and have obtained a degree of information corresponding to these requirements.[70]

Great strides were made within the general population in basic literacy, but perhaps the greatest impact on education was changes to the curriculum, structure, and methodology found within the universities. Despite each institution's small size, Scotland's universities in Edinburgh, Glasgow, St. Andrews, and Aberdeen had become prominent international centers of learning, and were in the forefront of the new thinking brought forward with the Enlightenment. (**Figure 2.3**) In addition, in contrast to the settings of their English counterparts, Oxford and Cambridge, isolated from the metropolitan areas of England, the urban locations of the Scottish universities allowed both students and faculty members to experience all aspects of city life, and provided them with opportunities for economic growth.

---

[*] In that era grammar schools were attended by boys whose parents were able to pay fees, though these would often be subsidized by the local burgh; places were also given free of charge for promising boys from poor families. At the start of the 18th century only Latin and Greek were taught, but over the years English, arithmetic, mathematics, and geography were introduced.

[†] scholarships

Figure 2.3 Old College Glasgow (Glasgow University)

It was also a significant social venue for men of letters, in that bankers, merchants, and clergy all mingled with the students and their instructors in the intellectual and social mix provided by the cities. These factors positioned the university communities at the vanguard of the Enlightenment.

Perhaps the greatest change occurred in the way individual courses were taught. In the late 17th and early 18th centuries in Great Britain university instructors, known as regents, made their living largely through the collection of student fees paid by each student enrolled in a particular course, and the students were assigned to a single regent, who would teach all their classes through their entire university program. But in 1708 — well ahead of the other universities —the University of Edinburgh abandoned the regent system. This eliminated the direct payment to regents and moved to a teaching system which allowed professors to teach specifically in their subject specialties across all divisions within the university. This exposed students to numerous professors during their course of studies, and improved the overall quality of the programs, as professors taught in areas they knew best. This subject-specific, lecture-based curriculum allowed subjects to be taught in greater depth and encouraged observation and experimentation.[71] These changes allowed the Scottish universities to offer an outstanding high-quality broad-based education. It allowed them to become internationally recognized, which in turn helped to put Scotland at the vanguard of new thinking and new ideas.

At the time, access to the Scottish universities was more open than in England, Germany, or France. This was especially true of the University of Edinburgh, uniquely founded by the town council and not by a religious order. From the outset, the University of Edinburgh accepted students of all faiths, whereas in England and Ireland only members of the Church of England could attend Oxford, Cambridge, or Trinity College Dublin. Catholics would not be allowed to attend Trinity until 1793 and would not be allowed at Oxford or Cambridge until 1871.[72]

In addition, universities in Scotland also had none of the class restrictions seen below their southern border; instead, they accepted students of all social and economic classes. The tuition in Scotland was £5 per year, one-tenth of the cost of attending either of the English universities. The less expensive tuition, along with the system of universal pre-collegiate education already existing throughout the Scottish Lowlands, meant that the graduates from universities in Scotland were more socially diverse and were much more likely to include students from working and middle-class families as contrasted with the top-heavy landed gentry and aristocratic student bodies found in the English programs.[73]

By 1750 Scotland's universities had progressed from being small and parochial institutions, largely for the training of clergy and lawyers, to major intellectual centers at the forefront of Scottish identity and life.[74] The language of the lectures at the Scottish universities was changed from Latin to English. Observatories were built at St. Andrews and Aberdeen. In all four universities, chairs were established in mathematics, law, and medicine. The Enlightenment saw a continuation of strengthening and improving programs and facilities within these already successful and growing Scottish universities. Edinburgh Town Council developed a policy of enlightened self-interest and promoted the city by helping to build the prestige of the university's academic, medical, and scientific life. The council regarded the university in general, and the medical school specifically, as institutions which if given enough prestige would not only stop the drift of Scottish students to foreign universities but also attract fee-paying students to Edinburgh from across Europe and America.[75] A working-class background, coupled with a first-rate university education, produced Scotland's uniquely qualified surgeons and physicians, destined to become the medical explorers.

A well-educated society not only provided strong schools and well-prepared teachers, but also gave rise to other changes. A dramatic increase in the number of newspapers, publishing houses, and libraries took place. Between 1760 and 1790 the number of newspapers in Edinburgh tripled. In 1763 Edinburgh had six printing houses and three paper mills. By 1783 there were sixteen printing houses and twelve paper mills, and most households, even those of lesser financial means, had a collection of books. What could not be purchased could be found at the local lending library; the first circulating library in Britain was established in Edinburgh in 1728, and by 1750 virtually every town of any size enjoyed this benefit. The American industrialist Andrew Carnegie, having been raised in Scotland, was an early benefactor of both its free schools and its circulating libraries. Among his many philanthropic efforts was the establishment of public libraries throughout the English-speaking world, and he opened the first Carnegie Library in his hometown of Dunfermline.[76] An appreciation of literacy, and innovation in pedagogy from the primary schools through the universities, prepared and supported a foundation for the movement that became the Scottish Enlightenment, and ensured that Scotland would become Europe's first modern literate society.[77]

Clubs and societies

In France, the Enlightenment was based in the salons where discussion and debate resulted in the published contributions of writers and philosophers, culminating in the creation of the

Encyclopédie. At the heart of the Scottish Enlightenment were the churchmen, lawyers, and professors who in the course of their daily lives relied on the spoken word in sermons, in the courts, and in the classrooms. On the Continent, discussions led to publications. Although the Scottish Enlightenment too was rich in publications, the soul of the movement was formed around the art of conversation and oratory, and this focus gave birth to a collection of new— and very Scottish—institutions: the clubs and societies. This uniquely Scottish network played a major role in the further development of the intellectual life associated with the Scottish Enlightenment.[78]

The milieu of the clubs and societies was urban, and during the Enlightenment the cities provided the perfect venue to allow the intelligentsia to meet regularly and casually. Because of the metropolitan nature of the Scottish Enlightenment, lawyers, academics, and clergy were joined by businessmen, city leaders, architects, wealthy landowners, and others in intimate settings that were convivial and social. One of the crucial features of the clubs and societies was the interconnection and relationships built between classes and disciplines within this distinctively Scottish setting. So successful and popular were they that during the Enlightenment dozens of clubs and societies were formed.[79] Some were short-lived and created only for dining and drinking. But others were designed for specific scientific bodies that still exist today, such as the Royal Society of Edinburgh. They also included social clubs open to men, regardless of their occupations, who simply enjoyed meeting and discussing a variety of topics across an intellectual continuum. In 1755 the mission of the clubs and societies was articulated in *The Scots Magazine*:

> The intention of these gentlemen was, by practice to improve themselves
> in reasoning and eloquence, and by the freedom of debate, to discover
> the most effectual methods of promoting the good of the country.

The clubs and societies were not limited to Edinburgh. Glasgow's earliest entry appeared when the university provost, Andrew Cochrane, founded the Political Economy Club. Its members met weekly, with the aim of creating links between academics and those outside the university setting. Its membership was drawn largely from the city merchants, though economist Adam Smith was also a member. The club's objective was "to enquire into the nature and principles of trade in all its branches, and to communicate their knowledge and views on that subject to each other." Smith indicated that he found the club invaluable to him in collecting material for *The Wealth of Nations*.[80] The debates within the societies and clubs encouraged young men to be "keen doubters," and to be effective speakers, as they learned to "talk the sun down." [81] The members of the intellectual elite, or "literati" as they were known, found these groups to be the focal point for the exchange of ideas, and positioned the clubs and societies at the forefront of the Scottish Enlightenment.

Edinburgh emerges

It can be said that Scotland's Age of Enlightenment was a product of economic growth brought on by the combination of several factors: the merger with England; the support of a

Figure 2.4 City View Edinburgh 1765 (Yale Center for British Art/PD)

structured church-led society firmly in place for over a century; the backing from the societal and financial elite of the law profession; and the already educationally advanced population, improved even more through major advancements in general literacy and a revamping of the university system. Because of these developments, a society emerged that was sympathetic to knowledge and not afraid of change.

The collaboration between the Church, the legal profession and the universities was present throughout the Scottish Lowlands, but the essence of the Scottish Enlightenment was contained within the city of Edinburgh. **(Figure 2.4)** Edinburgh is the historic capital of Scotland, and the University of Edinburgh had become the scholastic center of Europe and "a true reflection of the Scottish Enlightenment." The city, the seat of Scotland's Supreme Court and the General Assembly of the Kirk, quickly became the leader of the revolution. It would not be an exaggeration to state that Edinburgh was the heart and soul of Scotland and the Scottish people.[82]

But by 1815 although the Scottish Enlightenment was waning, the British Imperial Century was on the rise. The Enlightenment had created an energy that produced economic, social, and educational growth that placed Scotland at the forefront of European development. Although the Enlightenment had come slowly to Scotland, when it did come it reached such a flowering that Edinburgh was called "the Athens of the North." [83] It was from this revitalized and energized society that the newly qualified doctors would reject the life of a practicing local physician. And from that group would emerge an extraordinary coterie of daring surgeons, explorers, and naturalists.

3

A Somewhat Quiet Revolution

What you need to learn how to do is analyze situations and do differential
diagnoses and understand the principle and the concepts
rather than learn all the details,
and most medical schools don't begin to do that.

–LeRoy Hood

The Scottish Enlightenment produced the social, economic, cultural, and intellectual framework that positioned Scotland as the European nation in the forefront of the world's academic advancement and scientific thought. Under the influence of that framework, the University of Edinburgh's School of Medicine and its preparation of doctors would become the standard for the entire world. Understanding the evolution and progression of Scottish medical education is a critical component in comprehending why Scotland in general, and Edinburgh in particular, were so well positioned to produce the medical explorers who would lead the British Imperial Century.

Barbers and surgeons

During the early Middle Ages the practice of medicine throughout Europe had been dominated by the overwhelming influence of the Roman Catholic Church. During that period, Catholicism had dominated all aspects of society; for several centuries medicine was practiced within the walls—spiritual and physical—of the Church, and the medical discourse was largely based on the dialogue of religion.[84] Care of the sick and injured in Scotland was provided by the monks and friars who administered the hospitals connected to abbeys or priories, and dispensed health care as part of the general hospitality offered by these religious houses. Abbeys also created gardens where the monks grew a variety of plants to be used in medical treatments. But by the early 12th century the Catholic Church had begun to question whether surgery was an occupation that should be carried out by its clerics.

In 1163 a papal edict, *ecclesiac abhorret a sanguine*, issued by Pope Alexander III created a systemic change.[85] In essence the edict stated, "Shedding of blood is incompatible with

26

the cleric's holy duty to God," meaning that monks were no longer permitted to perform procedures such as surgery or phlebotomy.* Looking for a workable solution, the clerics enlisted barbers to assist them in the care of the sick and injured.[86] By assigning responsibilities to the barbers for procedures where contact with blood might occur, the monks could continue to operate the hospitals while maintaining compliance with the papal edict. The barbers, in addition to cutting hair and giving shaves, would now deliver a variety of medical procedures, and they became known as "barber surgeons." These men pulled teeth, performed surgery, and conducted bloodletting. But they also practiced medical interventions such as lancing boils, excising cysts and tumors, setting broken bones, and treating wounds. **(Figure 3.1)** This arrangement between the members of the clergy and barbers continued basically unchanged

Figure 3.1 Barber Surgeons (Wikimedia Commons/PD)

for the next 300 years, at which point the informal agreement between the two groups was formalized.

In Scotland from the late 15th century, groups of tradesmen of various sorts began to organize themselves into craft incorporations. Edinburgh was a typical medieval burgh, and its political and economic life came to be dominated by merchant and trade guilds. Individual artisans were still free to practice their skills, but tradesmen whose profession was not protected through incorporation were at an extreme disadvantage.[87] For the barber surgeons, the next step in becoming recognized as a legitimate profession meant incorporating themselves as a recognized trade. In 1505 the Town Council of Edinburgh granted a Charter of Incorporation, and the barber surgeons were formerly recognized as an "incorporated trade" within the city and surrounding area. The charter was ratified the following year by King James IV, making the Incorporation of Surgeons and Barbers of Edinburgh the first professional medical organization in Scotland.[88]

During the 18th century in Edinburgh, as in other cities throughout the British Isles, a bitter rivalry between the surgeons and the physicians would emerge. But this would be well in the future, when physicians emerged from academia and challenged the surgeons for leadership of the Edinburgh medical community. For the present, the Incorporation of

---

* Opening the veins for bloodletting.

Surgeons and Barbers would reign supreme. And as the first recognized medical organization in Scotland, it predated the founding of the University of Edinburgh by nearly 80 years, and the establishment of the university's medical school by over two centuries.

The University of Edinburgh

Founded in 1582 the University of Edinburgh, like the Incorporation of Surgeons and Barbers, was established by Edinburgh Town Council and was not affiliated with any religious organization. As mentioned earlier, this made the University of Edinburgh significantly different from England's Oxford and Cambridge universities, and Ireland's Trinity College, as each of these institutions was affiliated with the Church of England and would only accept male students of that faith. **(Figure 3.2)** But Edinburgh also differed from the other Scottish universities, at St. Andrews (founded in 1413), Glasgow (1454) and Aberdeen (1496), in that these institutions, having originally been founded by Roman Catholic bishops through papal authority, their student bodies were limited to male members of the Catholic faith. Then the Reformation in 1560 saw those three universities convert to Protestant institutions and limit enrollment to male students practicing that belief. The University of Edinburgh, however, although closely tied to the Dutch Protestant universities (especially Leiden), accepted students of all faiths from its inception. This lack of religious affiliation had many advantages in times of religious conflict, and also allowed easier entry for students from the English colonies and throughout Europe. But without a church's financial support, this also created a distinct disadvantage, as the university was often underfunded.[89]

Figure 3.2 University of Edinburgh
(Wikimedia Commons/PD)

The University of Edinburgh was modeled on the University of Bologna, and as in that institution the three most popular university courses of study were religion, law, and medicine. Religion required dedication and a calling to service, and although serving God was often the selected career path for sons of the working class, it was not seen as an occupation that provided either financial or social upward mobility. Law, once it was offered as a course of study within the Scottish universities, was a far less expensive proposition than it had been when its study had required attending a continental university. But it was still a career for the affluent. Those electing to study law were almost exclusively children of the wealthy. Thus, with religion drawing the sons of the working class and law attracting the offspring of the financial upper class, many of the well-educated sons of middle-class Scotsmen entered the universities to take the only remaining option: that of physician.

Surgeons, physicians and apothecaries

The Surgeons and Barbers, like the incorporated groups that had preceded them, trained their members by means of apprenticeship. The apprentice surgeon worked under the direction of a master surgeon who was required to "knaw the nature and substance of everything that he werkes, or ellis he is negligent." [90] A license to practice was awarded to the apprentice after he had completed a final examination, and so the graduates of these programs were designated Licentiates.[91] Those intending to become a Member of the Incorporation of Surgeons and Barbers pledged themselves as an apprentice to a master surgeon for a period of five years. Apprentices who did not intend to become Members of the Incorporation, and were simply looking for basic surgical training, generally served a three-year apprenticeship.[92] But then from 1778 the incorporation also issued licenses for students who had not completed the three-year apprenticeship, indicating that the student was "sufficiently qualified" to act as a surgeon's mate in His Majesty's Service.[93] In effect, this meant that surgeons attending to members of the military did not need to be as highly qualified as those treating the general public. The license allowed them to enter the navy or army immediately, earn a decent living, and perhaps return to Edinburgh at some time in the future to complete their medical studies, which many of them did.

Initially, the incorporation experienced few difficulties in attracting apprentices into its three program options, and the limited records from the time indicate that the affiliation between the barbers and the surgeons initially worked well. However, the organization's meeting minutes near the end of the 16th century show a rift beginning to open. By 1589 the disagreements had grown too great for the town council to ignore, and an ordinance was passed stating that barbers would continue to be admitted into the incorporation but were "forbidden from the practice of chirurgie* under pain of loss of their freedom." The Incorporation of Surgeons and Barbers continued, but each side had now evolved into the practices that their names implied.

In 1645 the Incorporation of Surgeons and Barbers admitted two apothecaries, James Borthwick and Thomas Kincaid, into their membership as "apothecary surgeons." James

---

* surgery

Borthwick was appointed "operator" for the incorporation and became its first regular teacher of anatomy. At about the time that the two apothecary surgeons were admitted, the incorporation purchased land and built a "physic garden," whose main purpose was to cultivate plants for medical use. Beginning in 1656, Borthwick and Kincaid used this early botanic garden to teach the "art of pharmacie" to the incorporation's apprentices. The discipline of pharmaceutical science was thus conceived and established in Edinburgh.[94] The Surgeons and Barbers, extending their power over the city's apothecaries, formed the Incorporation of the Surgeon Apothecaries in 1657. For all practical purposes, the association between the surgeons and barbers ended when the surgeons merged with the apothecaries. From that point forward, the minutes from the incorporation addressed only surgical training and medical matters, with no mention at all of the barbers.[95]

Although the University of Edinburgh did not have a medical school until 1726, its students could attain a medical diploma through their general studies coursework, by observing surgical techniques offered by the incorporation, and by completing courses in anatomy and the natural sciences. Because the Incorporation of Surgeons and the University of Edinburgh were both under the supervision of the town council, university students could observe anatomy demonstrations within the incorporation. Those students could then apply to take a written examination and, if successful, be granted a degree in medicine by the university.

The ability to earn a degree, as opposed to obtaining a license, was much more about status than educational preparation. No one could argue that the surgical training lacked academic rigor, but the social status ascribed to the surgeons as opposed to the physicians could not have been more different. Surgeons were regarded as inferior to physicians in all aspects, socially as well as occupationally; the university-educated physician was recognized as a gentleman, while the apprenticeship made the surgeon a tradesman. It was said that when visiting the home of a patient the physician would enter through the front door while the surgeon was required to use the servants' entrance.[96] This treatment was not appreciated by the surgeons, especially since their organization had predated the founding of the university by decades, and the surgeons constantly sought the equality they felt they deserved.

The Edinburgh surgeons were not unique in their attempts to improve their image and to place their ranks on a more equal footing with physicians; the movement was taking place throughout Europe during that period.[97] In surgical incorporations throughout England and on the Continent, classes in anatomy, surgical technique, and other science-based courses were being combined with a system that had previously been based exclusively on an apprenticeship. But beyond Edinburgh, surgeons were excluded from existing universities, and the incorporations had to develop separate educational institutions where they could prepare their apprentices. Edinburgh was unusual because all aspects of the city's medical programs fell under the guidance of the town council. This offered ease of access to the university for the surgical apprentices, and university lectures were just as available to them as to the university students. Thus, Edinburgh surgeons held a distinct advantage in their preparation when compared to their European counterparts.[98]

While the surgeons were attempting to elevate both their social and professional positions, the physicians were attempting to further distance themselves from those in the Surgeons'

Incorporation. One of the physicians' earliest efforts involved the application for a Royal Charter in 1617, but although the concept of a Royal College of Physicians was supported by the king, the move was opposed by leaders of the Church as well as by the Incorporation of Surgeons and Barbers. Two additional applications followed which were also unsuccessful. Finally, in 1681 the physicians were granted a Royal Charter by King Charles II, creating the Royal College of Physicians of Edinburgh (RCPE). Although the Incorporation of Surgeons and Barbers

Figure 3.3 Leiden University (Leiden University Library)

represented the oldest medical organization in Scotland, it was the physicians who were first recognized through a Royal Charter. This was another visible reminder of the difference in status, and further added to the surgeons' perception of a lack of respect.

The RCPE was created by highly trained, university-degreed medical practitioners who had earned their medical diplomas on the Continent, with 11 of the original 21 Fellows having graduated from the University of Leiden. **(Figure 3.3)** One of the more curious aspects of the agreement that had brought the RCPE into existence involved the negotiators agreeing that the college would not teach medicine; it did not have laboratories or medical classes, and it did not establish clinical teaching centers of its own. The founding physicians saw the RCPE as an organization that recognized medical excellence rather than one that provided resources to encourage it. Like modern professional groups, the Royal College of Physicians was most concerned with raising the status and visibility of its organization and its members, and in enhancing professional opportunities for its Fellows. The RCPE wanted to create an organization that would give preferential treatment to Scottish nationals graduating from either Scottish or continental medical schools. Ultimately the decision was made that graduates from any university with a degree as a medical doctor could be admitted to the RCPE following an examination, while graduates from Scottish universities could become Fellows without taking the exam.

Shortly after Borthwick and Kincaid had created the physic garden for the Incorporation of Surgeons, Sir Robert Sibbald and Sir Andrew Balfour developed the Edinburgh Botanic Garden for use by members of the RCPE. This physic garden, creating the base for the development of the study of *Materia medica** and chemistry within the University of Edinburgh, ultimately became the Royal Botanic Garden Edinburgh. While the University of Oxford's botanic garden (1621) had predated Edinburgh's by half a century, the way the gardens were used by each institution in the preparation of doctors varied greatly. While each utilized these resources to instruct doctors in the preparation of plant-based medications, the University of Edinburgh's botanic garden had become central to its required natural science

---

* pharmacology

curriculum decades ahead of similar steps taken by the English universities. The importance of the gardens to their respective universities can be seen by the status of the Botanic Garden Edinburgh's (BGE) curator within the hierarchy of the university. From the beginning, those responsible for its organization and development were not lowly lecturers in botany or pharmacy, as in other universities. The men trusted with the responsibility for directing the BGE were highly placed within the university, and their role extended to positions of importance within the greater Edinburgh medical community.

In 1685 Robert Sibbald was appointed the first professor of medicine within the University of Edinburgh. He was soon joined by Edinburgh physicians James Halket and Archibald Pitcairne within the university faculty. All three had been inspired by the innovative teaching style of the Dutch professor Herman Boerhaave, and were determined to introduce Boerhaave's methods into the study of medicine at the University of Edinburgh.[99] Application was made to the Edinburgh Town Council, which approved the incorporation of this methodology into the proposed program, and the three were appointed joint professors of the theory and practice of medicine at the University of Edinburgh.

Collaboration and growth

Despite the hard feelings created by the physicians being granted a Royal Charter, collaboration was seemingly the watchword within the City of Edinburgh's medical community. Apprentices within the incorporation were free to attend classes offered by the university, and university students often comprised the largest portion of the audience attending lectures and demonstrations conducted by the incorporation. Both groups were able to enroll in the extensive network of private "extra-mural" programs found throughout the city.[100] The extra-mural schools were private institutions, often focused on a single subject. Large numbers of students from both groups participated in their programs, which frequently combined talented teachers presenting courses not offered within either of the traditional programs, and which were often much less expensive than the incorporation's or the university's programs. Cooperation and collaboration became fundamental elements of the medical community's success. The closer alignment of the surgeons' program with the university, along with the surgeons' increased visibility within the community, also began to enhance their prestige.

Through these changes and innovations within the city of Edinburgh, something akin to the 21st-century medical practitioner began to evolve. In 1699 the surgeons built an "anatomical theatre," which was incorporated into what became known as Surgeons' Hall. **(Figure 3.4)** Lectures and dissections were conducted as part of the apprenticeship training but were open to university students as well. The lecture halls within the medical schools and extra-mural programs could at times find upwards of 500 fee-paying students in a single class; but while these students brought in a substantial amount of money for those conducting the demonstrations, this system ultimately created a serious problem for the profession.

The study of anatomy in the 18th and 19th centuries was, as indeed it is today, an essential part of medical training. Surgeons conducting the demonstrations needed subjects for dissection. But the use of human bodies for research and teaching was relatively new—and,

Figure 3.4 Surgeons' Hall (Wikimedia Commons/PD)

because of religious doctrine, was controversial. As the student audience grew, the demand for bodies exceeded the supply, and a black-market trade in corpses flourished. Within the city of Edinburgh a group of men, who would become known as "Resurrectionists," the most infamous being William Burke and William Hare, resorted to grave robbing and even murder to provide subjects for the surgeons. With so many students to teach, and so much money to be earned, many of the surgeons had no real interest in how or where the bodies came from.[101] But the trial and execution of Burke in 1828 brought the practice of the Resurrectionists into the full view of the public, and the impetus of this high-profile case forced the government to draw up the Anatomy Act in 1832. This at last provided for a tightly regulated, legal supply of cadavers for medical dissection, and ended the dubious trade between the surgeons and the Resurrectionists.[102]

The creation of the anatomical theatre had led to a Chair of Anatomy being added to the study of Theory and Practice of Medicine at the university. Dissections within the apprenticeship training, as well as those conducted as part of the public demonstrations, were conducted by surgeons from within the incorporation.[103] The addition of apothecaries, the opening of the anatomical theatre and the appointment of surgeons as faculty members within the university helped to make the barbers feel even more marginalized, and the resentment climaxed in 1722, when they formally separated from the incorporation by decree of the Court of Session and formed a separate guild. From that point on, barber surgeons ceased to exist.

The University of Edinburgh Medical School

Edinburgh continued to establish itself as the center for medical study in Europe and beyond, but while the Royal College of Physicians continued to evolve, the Incorporation of Surgeons occupied an increasingly more uncertain status. On the one hand, Edinburgh surgeons were prominent local practitioners. On the other, as the reputation of the University of Edinburgh spread, it was increasingly the professors of the Theory and Practice of Medicine who set the professional and educational standards for medicine and did so in the name of the RCPE. While it was entirely possible to practice as a physician in Edinburgh without becoming a Fellow, one's career would be more limited. So, despite the high costs of joining the RCPE, young physicians tended to do so, as it was regarded as almost mandatory for their personal and professional advancement.

The history of the University of Edinburgh Medical School is to a great extent the history of the Monro family. Anatomy was considered the cornerstone of a medical education, and the Monros ruled the anatomy dissecting rooms in Edinburgh for 28 years, from 1720 to 1748, with father, son and grandson holding the Chair of Anatomy in straight succession. As all three men were named Alexander Monro, they were referred to as "primus", "secundus" and "tertius". Monro (primus), the son of a military surgeon, had been born in London while his father was posted there. The family later moved to Edinburgh, where Monro (primus) served as an apprentice to his father, participated in anatomy dissections within the Incorporation of Surgeons, and attended classes at the university. Upon successful completion of his apprenticeship, he was sent to London to study anatomy. This was followed by study in Paris, where he performed operations under the direction of surgeons at the prestigious hospital, Hôtel-Dieu. He then travelled to Leiden to study under the tutelage of the distinguished Herman Boerhaave.

In 1719, Monro returned to Edinburgh and sat the examination to become a Fellow of the Incorporation of Surgeons. At the age of 22, with the support of this organization, the Edinburgh Town Council appointed him Professor of Anatomy at the University of Edinburgh. Monro, Sibbald, Halket and Pitcairne were soon joined by professors of botany, chemistry, *Materia medica*, and midwifery. In 1726, now with eight professors in place, the University of Edinburgh formally established its medical school.[104] Because of the continuous and extensive connections of the founding members to the medical school at the University of Leiden and their deep respect for Herman Boerhaave, it is not surprising that numerous elements of the Dutch medical program were soon major components in Edinburgh's newly founded medical school.[105]

The teaching hospital is born

Monro (primus) had long held the belief that a major element in replicating the Leiden model in Edinburgh would be the establishment of a teaching hospital. In 1721 he circulated a pamphlet among the citizens of Edinburgh setting out the case for what would become the Royal Infirmary of Edinburgh. His ally in this effort was George Drummond, the most powerful man in the city at the time; he had been elected Lord Provost of the City of Edinburgh no less than six times.* He supported the hospital project, as it could provide much-needed medical assistance to the poor of the city. However, his primary motivation was the prestige that the project would bring to him and to the city of Edinburgh if it could continue to be recognized as a center of medical innovation, practice, and excellence.

---

* The Lord Provost is the civic head of one of Scotland's principal cities: Aberdeen, Dundee, Edinburgh, and Glasgow.

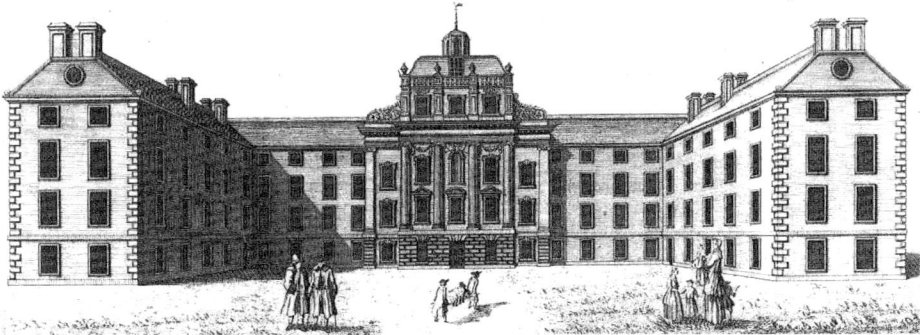

Figure 3.5 Old Royal Infirmary (Wikimedia Commons/PD)

Lord Provost Drummond worked with the College of Physicians to support Monro's appeal, and their efforts attracted £2,000 in financial backing from local surgeons, physicians, wealthy citizens, and the Church of Scotland parishes.[106] In 1729, Monro and the committee of donors established the Edinburgh Infirmary for the Sick Poor in a house on Robertson's Close, which was rented from the university. It was extremely small, with only six beds; because of this, and its location in a rented townhouse, it became known as the Little House. Here patients could be treated by students of both the Incorporation of Surgeons and the university, who could thus receive essential clinical training. During the 18th century education in medicine and surgery was greatly improved thanks in large part to the introduction of this innovative teaching hospital. In 1736 a Royal Charter was obtained, and the name was changed to the Royal Infirmary of Edinburgh. A guide for prospective students asserted, "The Infirmary of Edinburgh is much superior to any similar institution in Britain for the purpose of medical education." Treatments within the Royal Infirmary, reflecting both medical research and specialization, raised the medical school's reputation within the scientific and public communities, and the teaching hospital soon represented the pinnacle of medical achievement.[107]

The creation of a formal School of Medicine in 1726 and the establishment of the Royal Infirmary of Edinburgh in 1736 were in large part responsible for the University of Edinburgh becoming recognized as the center for the "systematic medical teaching established on a sound scientific basis." [108] With the rapid growth of the new medical school, Little House soon became too small, and it was replaced in 1741 by a 200-bed facility, the Old Infirmary. (**Figure 3.5**) This allowed the emphasis on practical, hospital-based teaching to become Edinburgh's hallmark, and gained the medical school an international reputation for excellence in clinical instruction.[109]

Medical societies

The formation of the clubs and societies during the Scottish Enlightenment had involved almost all of the Scottish literati, and the development of the medical profession in Scotland is best reflected in the rise of the medical societies. Like the other clubs and societies, their activities ranged from the educational through the professional to the social. In *Scottish Medical Societies 1731–1939: Their History and Records*, Jacqueline Jenkinson identifies seven types of medical societies:

- General Interest Medical Societies, which were open to all qualified practitioners and any branch of medicine, and any aspect of medical science could be considered for discussion.

- Convivial Societies, which were relegated to dining and social gatherings, and were designed to develop social contacts between members of the profession.

- Medico-Scientific Societies, which were created to share practical observations, communicate facts, and exchange opinions on medical subjects. Their discussions tended to connect medicine with subjects such as natural science, chemistry, or biology.

- Medical-Related Societies, which dealt with specific issues of strong medical interest such as medical missionary work or charity-based health societies, and were opened to lay people who could contribute to the non-medical side of the discussions.

- Single-Discipline Societies, Scottish Specialist Societies and Professional Protection Societies, each of which was devoted to a single medical discipline such as surgery, pathology, or dentistry.

But this last group of societies did not begin to appear until the end of the 19th century, and they followed the emergence of a separate class of medical "specialists" which occurred at the same time.[110]

The wider scientific and cultural advancement of Edinburgh over the other university cities during the Scottish Enlightenment is closely linked to the early emergence of the medical societies within the city and can be documented by noting the dates when these societies emerged. The first Scottish society was the Medical Society of Edinburgh, established in 1731, only five years after the formation of the University of Edinburgh Medical School. The focus of the society initially was to bridge the divisions between the physicians and surgeons, attempting to capitalize on their commonalities rather than emphasizing their differences. The Medical Society of Edinburgh was an indication of a new trend in Scottish medical practice. A key factor in Edinburgh's superior position among the university medical communities was the movement to improve the practice of medicine through scientific debate and a societal association between members of a common profession.

Jenkinson identified 137 separate medical societies that were created in Scotland between

1731 and 1939. However, Edinburgh's leadership in the creation of medical societies was challenged only towards the end of the 18th century, first by Aberdeen and then by Glasgow. The Aberdeen Philosophical Society, the first medical society in that city, began in 1758; this was 27 years after Edinburgh's first initiative. The Glasgow Medical Club began in 1798. Meanwhile, in the 67 years from the creation of the Edinburgh Medical Society, the city of Edinburgh had launched seven additional medical societies. These later appearances in Aberdeen and Glasgow were clearly based on the model of the Edinburgh clubs and societies.

It can be argued that the clubs and societies of Scotland provided the platform that led the Scottish Enlightenment, and that medical societies of Edinburgh in particular led to the development of the unique and highly successful system of medical education that came to characterize the city and its university.[111]

The differences disappear

While apprentices within the Incorporation of Surgeons continued to have an avenue for access to university courses, it was only professors who were Fellows of the Royal College of Physicians who could be designated faculty members within the medical school. Regulations within the Royal College of Physicians also prohibited its members from being members of the Incorporation of Surgeons. Edinburgh surgeons were thus prohibited from becoming teaching members of the medical faculty if they remained members of the incorporation, and the professional and social mobility provided by membership within the RCPE could only be achieved by leaving the privileges of the surgeon's guild behind.[112] However, the growing professional and public demand for improvement in the practice of medicine eventually resulted in a blurring of the traditional divisions between physician, surgeon, and apothecary.[113]

In 1778, when King George III granted a Royal Charter to the Incorporation of Surgeons and Apothecaries, it became the Royal College of Surgeons of the City of Edinburgh (RCSE). The charter gave the surgeons the professional status they had long been seeking while maintaining the privileges afforded them as a guild charted through the Edinburgh Town Council. With the granting of the surgeons' Royal Charter, the differences between the two royal colleges rapidly began to disappear. Now, surgical apprentices often outnumbered university students participating in university lectures. Surgery courses within Edinburgh University moved from being exclusively lecture based to clinically based demonstrations conducted by practicing surgeons. The teaching faculty working with the prospective surgeons and physicians each adjusted their regulations, and soon the similarities in the requirements for their respective graduates far surpassed the differences. In 1824 the medical school introduced a requirement for physicians to have three months' "practical experience" in surgery, replacing its long-standing program of lectures and demonstrations. Students from both programs gained access to the teaching wards in the infirmary through the use of a ticket which allowed them to attend clinical lectures and to observe surgeons as they treated patients.[114] Edinburgh became the leading location in Europe for the preparation of medical professionals.[115] Graduates quickly dominated the military medical services, and most outposts throughout the British colonies were filled by Edinburgh-trained doctors.

From good to great

One of the central pillars of the Scottish Enlightenment was the expansion of scientific and medical knowledge combined with the process of analyzing findings in new ways. This required medical professors who could see science through the eyes of utility, improvement, and reform. Through the brilliant research and teaching of individuals such as anatomist Alexander Monro, chemists William Cullen and Joseph Black, and natural historian John Walker, the University of Edinburgh not only became a major center of teaching but also was in the forefront of medical research. The work of these innovators stood as a model for others who shared their philosophy of science and education. By the end of the 18th century the University of Edinburgh Medical School, by then the dominant faculty within the university, was arguably one of the leading centers not just of medicine, but of the sciences throughout Europe.[116]

The University of Edinburgh's reputation as a preeminent, international center of learning grew, and the advances in medical studies were especially profound. Given their close geographic proximity it might be assumed that Scottish and English medical preparation, training, and practice would be somewhat parallel. But in fact, this was true only to an extremely limited extent; the preparation provided to Scottish medical students at the time was quite unlike that of their English counterparts.[117] In England, medical schools stressed caring and curing. In Scotland, however, the university's medical instruction, in addition to courses focusing on patient care, placed a significant emphasis on both the humanities and natural sciences. Chemistry and botany were part of the standard curriculum. Additional courses including meteorology, minerology, zoology, and geology were available to apprentices, medical students, and the public through the extra-mural schools.

In addition, the differences in the preparation of Scottish doctors extended beyond the expanded curriculum available to them. The Scottish medical programs were producing a new type of doctor which combined the roles of physician, surgeon, and apothecary. This necessitated a complete transformation in the way the medical professors and their students interacted. Initially, both Scottish and English students were taught through the didactic method, a teacher-centered method of instruction which was content-oriented, with the teacher giving instructions to the students and the students generally functioning as passive listeners. It became apparent, however, to the decision makers of the Scottish medical schools that learning the practice of medicine in a clinical setting required a method that was quite different, and the methodology evolved into what is now referred to as experiential learning. In order to accomplish this, the study of medicine in Scotland evolved by incorporating an emphasis on clinical diagnosis and system analysis; students learned from observation and hands-on practice rather than simply through lectures. In essence, the student learned through discovery rather than through the words of the teacher. This in turn taught students to look at individual patients as components of an overall system, and to question everything.

Medical achievements in Scotland during this period were substantial. Doctors used observation and experimentation and had access to patients in the teaching hospital. Scottish medical students were learning to describe, predict, and understand medical phenomena based on empirical evidence. The growth of laboratory science in assisting diagnosis and hospital-based clinical research further distanced the Scottish medical graduate from his English counterpart.[118] This new way of teaching and looking at the world would generate characteristics which would be fundamental to the genesis of the new medical explorers.

International impact

Well ahead of the British Imperial Century, Edinburgh was established as the seat of medical learning. Students from the far corners of the earth came to Edinburgh to complete their education. The city played host to thousands of students, not only in the medical and surgical schools but in other areas of advanced learning as well. To be truly educated was to have an advanced degree from Edinburgh.[119] By 1765 the fame of the Edinburgh School of Medicine had spread, and students were as attracted to Edinburgh as they had been to Leiden a generation before. This trend continued throughout the British Imperial Century, and by the beginning of the First World War the Edinburgh School of Medicine was widely considered the best in the world.

Between 1726 and 1799 no less than 195 students came from North America alone.[120] The School of Medicine attracted many Americans who went on to establish and teach at some of the first teaching hospitals and medical schools in their own country. In Colonial America, the first medical school, founded in 1765 at the University of Pennsylvania, was patterned after that of Edinburgh; to this day the Pennsylvania school's emblem remains the Scottish thistle. The founders, John Morgan and William Shippen, had both studied at the University of Edinburgh and four of Pennsylvania's first five medical professors had medical degrees from Edinburgh. The pattern followed with Columbia (1767), William and Mary (1779), Harvard (1782) and Dartmouth (1797), founded respectively by Samuel Bard, James McClurg, Benjamin Waterhouse, and Nathan Smith—all of whom had studied at Edinburgh.[121]

To sum up, Edinburgh's program, and those of the universities that modeled theirs upon it, implemented a progressive admission process where young men were selected to study medicine regardless of social status, economic background, or religious faith. These prospective doctors were prepared within a robust environment where they could participate in lively discussion, and where empiricism and inductive reasoning were the hallmarks of the debates. The use of a botanic garden as a teaching tool became an integral part of the medical program, allowing students to observe, test, and make connections between objects in the natural world. Students learned to see systems and interdependencies in nature rather than individual occurrences and learned to draw upon their knowledge of the natural sciences. Beyond the required courses, students could expand their interests in botany, zoology, geology, and other sciences through an expansive array of options offered within their programs. Finally, the inclusion of the Royal Infirmary combined a university education in humanities and natural science with the clinical study of patients in a hospital setting.[122] This innovative inclusion allowed students to interact with patients, ask questions, and monitor the successes and failures of their own work.

The practice of medicine in the real world created a dynamic and robust clinical experience that was unrivaled in any other educational establishment, and so a new generation of physicians began to emerge. These newly trained surgeons would become the medical explorers who would reject a comfortable medical practice, the economic and social status their profession could provide, and the pleasure of living among close friends and family, and instead would choose the lure of exploration of new and exotic lands. These men emerged from a culture that valued education for both rich and poor. They were the products of an extraordinary intellectual movement, the Scottish Enlightenment. The city of Edinburgh absorbed all these influences and produced a medical program not found in other universities. And the University of Edinburgh Medical School was graduating doctors who would use their skills and training in a new and unexpected way—as explorers.

4

The Military and the Chartered Companies

It was not, as some suggest,

Calvinism that made Scots hard:

it was the hard Scottish character that made Calvinism.

–James G. Leyburn

The Scots were endowed with what Leyburn calls "dourness," meaning hardness and durability, having the qualities of iron. Men who had survived centuries of living in a hard environment, both physically and socially, had learned how to endure. Scotland was a land of poverty, periodic famines, and a strong history of extensive migration long before the British Imperial Century, and for a Scotsman the opportunity to become an explorer in order to escape these hard conditions held much promise and little fear.

The proximity to, and reliance on, the sea was also an important factor. Scotland is a land of steep wooded slopes, long sea lochs, and rugged mountains. Before the coming of modern roads, overland travel within Scotland had been arduous and time-consuming, this meant that well into the 18th century the sea was the quickest and most convenient means of transport. The sea, sailors, and ships played an essential role in the lives of Scottish men, and this would be an important factor as Great Britain looked for men to explore and expand its empire.[123]

In addition, the Scots were educated men, and good schools were a prerequisite for good explorers. An uneducated man was at a disadvantage; the survival of the explorers, whether traveling in the heat of Africa or the cold of the Arctic, depended upon their knowledge as much as it did on their physical toughness. An explorer in the British Imperial Century had to be a man of learning and a man of action. He had to reach his destination, understand what he saw, and be able to complete an informative and useful report of his discoveries to submit when he returned home, so that others could then benefit from the knowledge.

An additional factor that helps to explain the high number of Scots who participated in the journeys of discovery was that Scotland was much poorer than England. Throughout the centuries there was a lack of employment options for those who remained in Scotland, even for those with university degrees. Economic distress and uncertainty about the future combined to drive many able and energetic Scots to leave their homeland and seek new lives and fortunes abroad. These were men who were much more likely to tolerate the hardships to be encountered in other parts of the world, and to accept the potential of financial reward offered through relocating to them.[124] And although options for travel had existed before the 18th century, the merger between the Kingdom of Scotland and the Kingdom of England dramatically increased these opportunities.

As mentioned earlier, when the Acts of Union in 1707 moved the socially and politically elite Scots from Edinburgh to London, this allowed those who remained—the lawyers, clerical leaders, and academics—to emerge and fill a void in leadership and to become the central players in what would become the Scottish Enlightenment. The Enlightenment would create in Scotland the most literate population in Europe. This would impact education throughout the country, and completely transform the university system and particularly the methodology used to prepare physicians. These were key factors in creating the medical explorers who would lead the British Imperial Century. But the Acts of Union also put into place the final measure that would allow the medical explorers to move to the forefront of the development of the British Empire. In addition to the military, Scots (and the medical men within it) could work within the English joint-stock companies such as the British East India Company, and Scotsmen could find employment within the scientific and cultural institutions located south of the border.

This meant that at the start of the Enlightenment the Scots were already established as an exceptionally mobile nation. In 1775 Edward Topham, an English journalist living in Edinburgh, stated:

> Go into whatever country you will, you will always find Scotchmen. They penetrate into every climate: you meet them in all the various departments of travellers, soldiers, merchants, adventurers and domestics.[125]

Scotland became a country whose principal export was its sons; leaving home became a traditional rite of passage. For those whose characters were forged in austerity and who could find no outlet for their industriousness at home, exploration was a natural choice. Their exodus was not always due to financial considerations; some left because of a spirit of excitement. This was an age of adventure, a time of exploring and settling new parts of the world, and these efforts required men of courage and ability to administer and police these far-flung lands. Adventurous men were needed to build up trade, to explore the hinterlands and clear routes through difficult terrain, to teach native populations, to bring Christianity to the "godless," and to cure the people of these far-flung colonies of illness and disease. The opportunities were endless, and the Scotsmen embraced them. During the Imperial Century, when Britain controlled nearly a quarter of the surface of the earth, Scotsmen predominated in every important sphere.[126] The new opportunities that emerged from the Acts of Union not only help to clarify why a Scotsman would leave his home and search for a better life, but also explain why so many did so in the service of England.

The British military

Scotland had maintained its own army and navy for decades, and continued to do so into the reign of Charles I. Service in the military, where units tended to be formed from among men of a specific region (or even from a single landowner's estate) allowed for unity while observing the structures of social class represented by the system of military ranks.[127] But after Cromwell's victory in 1651 the Scottish army was disbanded, and the Royal Scots Navy had its ships and its crews divided among the Commonwealth of England's fleet; both the ships and the sailors were forced to serve under English captains. Upon becoming part of the Commonwealth, Scottish merchant seamen received protection against impressment.* However, each of the seacoast burghs was assigned a fixed quota of annual conscripts for the English Royal Navy. This meant that between Scots volunteers, looking for a way out of poverty, and the annual collection of conscripts, the English Commonwealth Navy soon became disproportionally manned by Scotsmen.

When Charles II assumed the throne in 1660 Scotland had a navy, but no ships beyond privateers hired on an as-needed basis. His Privy Council recommended the reestablishment of a permanent navy and army for Scotland, and the Scottish Army and Navy were placed under the control of the monarch, who expanded both branches. The ground troops were soon a formidable force, and Scotland could then boast a standing army of seven units of infantry and two of horse, together with various levels of fortress artillery. In 1697 it was decided that Scotland needed to reestablish a professional navy for the protection of commerce in its home waters. English shipbuilders provided three purpose-built warships: the *Royal William*, the *Royal Mary*, and *Dumbarton Castle*. But this fledgling navy was to exist for only a brief ten years.

The Acts of Union in 1707 presented the men of Scotland with two military options as avenues for gainful employment. The Scottish office of Lord High Admiral was subsumed within the Admiralty Office of Great Britain, and the Royal Scots Navy was merged with the British Royal Navy, so the vast majority of Scottish naval officers and crewmen simply changed into the uniform of the newly combined force.[128]

The Scottish Army, like the Navy, was combined under one operational command. However, its amalgamation with the British Army required even less of a transition, as Scottish regiments, including the Scots Guards, the Scots Greys, and the Royal Scots of Foot, were simply transplanted as intact units into the British forces. Although all the regiments were now part of the new British military establishment, they remained under the old operational command structure, and retained much of the institutional ethos, customs, and traditions created in Scotland years earlier. **(Figure 4.1)** Following the signing of the Acts of Union, enlisting within the British Army or the Royal Navy became a favored choice for Scotsmen looking for a better life. The number of Scotsmen selecting the British military as employment continued to dramatically increase, and while at the end of the 18th century the Scots constituted less than 15 percent of the British population, the number of Scots serving in the British military represented 36 percent of the volunteer corps of the army and navy.[129]

---

* Taking men into the military against their will.

Figure 4.1 92nd Gordon Highlanders (Wikimedia Commons/PD)

By the time of the Napoleonic Wars in the early 19th century, the number of Scottish recruits was unmatched by the Englishmen. The most careful estimate suggests totals ranging from 37,000 to 48,000 Scotsmen in regular, fencible,* and volunteer units.[130]

Not all the Scotsmen who elected to enter the military did so as fighting troops. As mentioned earlier, because Scotland had the most advanced system of universal education in Europe, large numbers of well-educated Scottish men entered university to study medicine, and those who chose to become doctors generally came from middle-class Scottish families. Upon graduation from medical school, these sons of the Scottish middle-class had the option to establish a medical practice in Scotland. But the Scottish universities were graduating far more doctors than Scotland could absorb. Therefore, large numbers of graduating Scottish physicians began to look to the military as a preferred career choice. The British military soon contained a disproportionally high percentage of Scottish doctors among their ranks; by the end of the 18th century most British military doctors were Scottish, and the majority of those were Edinburgh-trained.[131]

As Scottish culture is linked closely to the sea, many medical graduates became assistant surgeons in the Royal Navy. The medical graduates entering the army could look forward to a posting in a foreign land. This had clear advantages, including steady employment in far away, perhaps exotic settings. But advancement during peacetime in the army could prove difficult, whereas within the navy a much different set of criteria for promotion existed—a set of criteria that for the Scottish medical graduates looking to the military as a career the navy provided a distinct advantage over the army. During the British Imperial Century there were two primary ways to advance one's career in the navy: to distinguish oneself in battle or to serve as a naturalist on an exploring voyage. Between 1689 and 1815 the British were almost

---

* Troops stationed to protect the homeland or colonies from invasion, i.e. the Home Guard.

continuously at war, then for the next 100 years they largely avoided any European military entanglements,[132] and in this period of relative peace it was exploration that became the key to advancement for naval doctors.

The doctors from Edinburgh, with their background in natural science, botany in particular, were ideally suited to perform dual roles on exploring voyages.[133] They could serve the needs of the crew as a surgeon, but also their significant knowledge of natural science allowed them to collect, examine, analyze, and document the plants, animals, and artifacts encountered on these voyages. Serving as the naturalist and chronicler of the expedition also provided an opportunity for recognition and personal fame. Being far better educated than the other ship's officers, and with their strong scientific skills, these medical men could be expected to record their observations in a systematic manner and they were, in addition, often also accomplished in drawing and painting. Upon returning home they were likewise best suited for converting the journals detailing these voyages into published narratives. The rank of assistant surgeon was a high one which paid well, and on advancement from assistant surgeon to surgeon, the salary was usually only exceeded by that of the captain.[134]

Graduates of the University of Edinburgh were in high demand, and a connection between the Royal Navy and the university was soon well established. Eric Linklater, in the *Voyage of the Challenger* (1972), devotes a chapter to the connection that had been established between the Royal Navy and the University of Edinburgh by the middle of the 19th century, and the role that they jointly played in advancing the empire. He states:

> Where praise is due – and praise is due in plenty – it may be that priority should be given to the Royal Navy, but close behind … was the University of Edinburgh which since the second half of the eighteenth century had been expanding in virtue and increasing its influence. Its reputation for scholarship was generously based on its Medical School … and in the early years of its progress towards a continental fame its teaching was acclaimed far beyond that of Oxford or Cambridge.[135]

Through scientific interest, the adventure of travel, and perhaps a lack of other options, these newly graduated physicians turned in large numbers to the Royal Navy. Throughout the British Imperial Century most of the important voyages of exploration had Scottish-trained surgeons among the expedition's members. During this momentous era the doctors of Edinburgh were uniquely prepared, ideally placed, and eager to serve on the exploring voyages.

The joint-stock companies

Timothy Parsons, in *The British Imperial Century, 1815–1914: A World History Perspective*, defines "empire" as a word dating from Roman times, when an *imperium* was the territory ruled by an *imperator*, or general. Empires were generally built through military conquest. But in fact, much of Britain's imperial territory consisted of colonies settled by emigrants from Great Britain, and through formal and informal alliances created through trade. Parsons states that a better definition of the British Imperial Century would be the era in which a previously "informal" empire based on trade and commerce was transformed into a "formal" empire based on territorial acquisition or at least regional control.[136] With the Acts of Union, employment prospects had arisen within the completely reformed British Military. But union with England had provided additional entities offering options for Scottish employment in international settings; it offered opportunities for the uniquely prepared medical explorers to engage their skills and make their mark. Chief among these opportunities lay in the joint-stock company.

In 1588 the English defeat of the Spanish Armada had signaled England's earliest emergence as the dominant sea power around the globe, and the period from 1588 to 1783 became known as the First British Empire. With its empire expanding at a rapid rate, England began to look at creative ways to govern and protect vast areas of the globe while making those possessions financially viable.

Throughout the general course of history governments have turned to private traders to establish commerce and governance on their behalf. A charter is given to a trading concern, empowering it to perform the essential functions of government within a foreign country or territory and to pay itself out of the profits. Revenue generated beyond this amount reverts to the sponsoring government. These entities are called joint-stock companies.[137] This age-old system of the chartered company runs through the story of the British Empire during its three centuries from the time of Elizabeth I to the time of George V. In many cases, joint-stock companies were the primary agents of empire building. Operating under a Royal Charter from the British Crown, these private enterprises undertook the financial and human expense of opening markets and establishing settlements. In return they earned a monopoly on trade within their area of influence.[138] Most of these companies were short-lived, with over-optimistic investors underestimating the impacts of disease, harsh climate, unfriendly local populations, and the ruthlessness of their trading rivals. But some of the enterprises were successful, and although there were others, it was the East India Company and the Hudson's Bay Company which had the most significant influence on the lives of the medical explorers.

The East India Company

The English East India Company received its Royal Charter from Queen Elizabeth I in 1600. In its first 100 years the Company survived the fall of the House of Tudor, the introduction of the House of Stuart, the rise and fall of Cromwell, and the installation of William and Mary during the Glorious Revolution. But as the Company had been formed more than a century before the Acts of Union, it had initially employed few Scots. Although prior to the Union it had attracted a number of Scotland's landed families as investors, and there was a presence of Scots among its directors based in London, **(Figure 4.2)** its day-to-day operations in India were almost exclusively in the hands of the English. But when, on the merger of Scotland and England in 1707, the English East India Company became the British East India Company, giving Scotsmen much wider access to the rapidly forming British Empire. Even so, the English were slow to invite the Scots to join them on the subcontinent. But the creation of a standing army within the Company and the establishment of a new system for directing the Company would change all of that.

Until that point, the English, French, and Dutch each had their East India companies, all of them with outposts in close proximity to each other throughout India and much of Southeast Asia. With these three powerful nations all attempting to gain exclusive control of the region, conflict was inevitable. Parliament had earlier given the East India Company permission to resort to armed conflict if the need arose, and clashes between the English and French, between the English and the Dutch, and between the Dutch and the French, had been frequent. What became known as the Carnatic Wars (1744–1763), involving Britain and France tussling for control of India, were the largest of these conflicts. Throughout the region the independent local rulers aligned themselves with, and provided military forces for, either the French or the British. In India, after nearly 20 years of continuous warfare, the local rulers supporting the British East India Company were victorious.

Figure 4.2 East India Company Headquarters, London (Wikimedia Commons/PD)

As a result of these battles, the British established almost total dominance within India, but the British East India Company realized that its reliance on military support provided by the local rulers in order to defeat the French pointed to the necessity for the Company to establish its own army. Moving rapidly, it created a private army of about 260,000 troops, twice the size of the British Army; it was service in this military unit that provided the initial major inroad for Scottish employment. The sepoys* outnumbered the British troops at a ratio in excess of 10 to 1. However, the officers and non-commissioned officers were recruited exclusively from the populations throughout Great Britain, and hundreds of Scotsmen rushed to join the army's ranks; indeed, of the 116 candidates for the officer corps, 56 were Scots. But the large numbers of transplanted Scotsmen who would seek employment within the East India Company were not restricted to the army.

With the military success in the Carnatic Wars, and the establishment of a standing army, the internal structure of the East India Company changed from a focus on trade to one of territorial expansion. But for the government in London, it was not happening quickly enough. In 1783 William Pitt the Younger declared publicly that he was unhappy in that after nearly 200 years the English monopoly created through the Company had failed to gain total domination over India and the surrounding region. Parliament's solution was to pass the 1784 India Act, whose purpose was to give the British Government more direct control over the Company's dealings. While its board of directors maintained their ability to conduct the day-to-day business of trade, Prime Minister Pitt announced that he was creating a Board of Control for Indian Affairs, which would be the overall governing force for the Company.

---

* Indian troops

This body would come to be dominated by Pitt's good friend, Scotsman Henry Dundas, who served as president of the board and effectively ran the Company until 1811. He was also responsible for encouraging large numbers of his countrymen to look to India for employment. Soon Scotsmen become the prevailing work force, and quickly dominated all aspects of the Company.[139] Scots came to the company as writers, factors, traders, engineers, missionaries, indigo planters, jute traders, and teachers. The Scots also came in large numbers as company doctors and as ships' surgeons on the company's trading vessels. In 1800, of the 254 assistant surgeons recruited to the Company, 132 were Scots.[140]

These doctors were all graduates of Scottish universities and, as within the medical explorers as a whole, most were University of Edinburgh trained. For many of the medical explorers, such as Mungo Park or William Jardine, their service for the Company provided their initial experience as explorers, as their medical duties were closely linked to the establishment of the Company's trade routes, the charting of newly incorporated lands, and the establishment of the empire's expanded financial base. Most of the botanical gardens that became the pillar of what became known as Economic Botany were established within territory controlled by the Company and were managed by the Scottish medical explorers in their employ.

During its heyday the Company conquered territory, established governmental institutions, created armies and navies, and even produced its own currency. At the height of its rule the British East India Company controlled approximately two-thirds of the world's trade. But the Indian Rebellion of 1857 finally brought it down. The Company was initially caught off guard, and it was months before its officers could recover and put down the revolt. They ultimately did so only with the help of a large number of sepoys who had remained loyal to their British masters.

Parliament had been looking for a way to end the company's control, and as soon as the insurrection had been suppressed, it saw an opportunity and took it. In August 1858 Parliament, at the direction of Prime Minister Disraeli, issued the Government of India Act, which demanded that the company's officials transfer all their power to the Crown. The British East India Company had ruled India in the name of Great Britain for almost 260 years, but with the passage of that Act the British Government assumed direct control, in the form of the British Raj, and the British East India Company ceased to exist.[141]

The Hudson's Bay Company

In 1670 the Hudson's Bay Company was granted a Royal Charter by Charles II, and this joint-stock company became to North America what the East India Company was to Asia. The granting of the charter had a threefold purpose. First, the Company was to occupy lands adjacent to Hudson Bay in the name of England. Second, it was to carry on any commerce within those lands that might prove profitable. Third, in the same way that explorers connected with the East India Company were charged with establishing trade routes throughout the Indian Ocean, a major focus for explorers working within the Hudson's Bay Company was to search for the Northwest Passage, a sea route that would connect the Atlantic and Pacific Oceans around or through North America. The charter granted the Company a monopoly of the region drained by all rivers and streams flowing into Hudson Bay. This drainage basin spans 1,490,900 square miles, comprising over one-third of the area of modern-day Canada, and stretches into present-day North Central United States.[142]

But there was a problem with this charter; the land in question, a vast region, had been claimed by France nearly 200 years earlier and was known as New France. For much of the 17th century and into the early 18th the French had a monopoly on the fur trade throughout North America, and an entire network of middlemen, exporters, and importers became rich on the efforts of the trappers and their native allies. In the 1630s two French traders, Pierre-Esprit Radisson and Médard des Groseilliers, learned that even better prospects lay in the unexplored territories around Hudson Bay, and they sought French backing for a plan to establish a trading post in the region. These two entrepreneurs approached the French governor and proposed mounting an expedition to scout the area in an effort to greatly expand the French fur trade—but the governor refused permission for their travel. Still determined to establish their trade in the Hudson Bay area, the two Frenchmen approached a group of English merchants in Boston to help finance their explorations, and in 1665 the merchants brought the pair to England, where they were fortunate enough to meet Prince Rupert of the Rhine, a close cousin of King Charles II.

The English were looking for new revenue streams, and if they could establish an enterprise near their already growing American colonies, the opportunity would be logistically and financially ideal. If they could harm or disrupt their long-time rival, so much the better. Charles approved the charter, and the territories were granted to the Hudson's Bay Company. The area became known as Rupert's Land after Prince Rupert, who the king had appointed the first governor of the Company and who had originally helped broker the deal. Thus, two Frenchmen received a charter from the King of England and established an English joint-stock company in the middle of New France.[143] But then over the next 30 years France and England fought for control of this Canadian territory. Finally, it was settled by wars in Europe rather than in North America; because of England's success in the Nine Years' War in 1697, and the War of the Spanish Succession in 1713, the Treaty of Utrecht was signed. This document required France to relinquish all claims to Great Britain related to the Hudson's Bay Company, and for the first time the lands occupied by the Company became an undisputed British possession. As with the East India Company, the focus of the Hudson's Bay Company moved during the 18th century from trade to territorial acquisition; it continued to absorb new territory, and eventually controlled almost all of present-day Canada.

In early Canada, the fur trade was by far the dominant economic activity, but by the time the Scots began arriving in the early 18th century it had already been well established by the French in Quebec and the English in the area controlled by the Company. However, the Scots quickly caught up; by 1799 four out of every five employees of the Company were Scots, and the majority were from the Orkney Islands.[144] Early in the 18th century, ships outbound from Britain for the Company were in the habit of stopping for supplies in the Orkneys, and they soon began recruiting Orcadians for work in Rupert's Land. These recruits proved themselves to be hardy, hardworking, and industrious. When the company expanded its workforce from 180 to over 600, three-quarters of the population were Orcadians. These Orkney Islanders joined large numbers of workers from other parts of Scotland, and by the end of the 18th century Scotsmen so dominated the trade that Elaine Allan Mitchell says:

> It would be almost impossible to over-emphasize the pre-eminent position which Scots of every stripe, Highlander, Lowlander and Islander, attained during the eighteenth and nineteenth centuries in the North American fur trade.[145]

In 1867 the Canadian Confederation united the three colonies of Canada, Nova Scotia, and New Brunswick into the Dominion of Canada. The Hudson's Bay Company continued to exist until 1869, when it was directed by the British Government to sell its land to the newly formed Canadian Government in what is known as the Deed of Surrender. By that time the Company had become the largest corporate landowner in history, its realm encompassing nearly 3 million square miles. **(Figure 4.3)** By 1869, when it sold its land to the newly formed Government of Canada, it is estimated that out of the company's 263 commissioned officers 171 (65 percent) were Scotsmen.[146]

Creating an exploring base

Service in the British Army or the Royal Navy was often the entry point for the medical explorers, whether they were stationed in foreign lands or traveling the seas during remote assignments to chart new lands or to find hitherto undiscovered plants or animals. The same was true when accepting positions within the joint-stock companies. The East India and

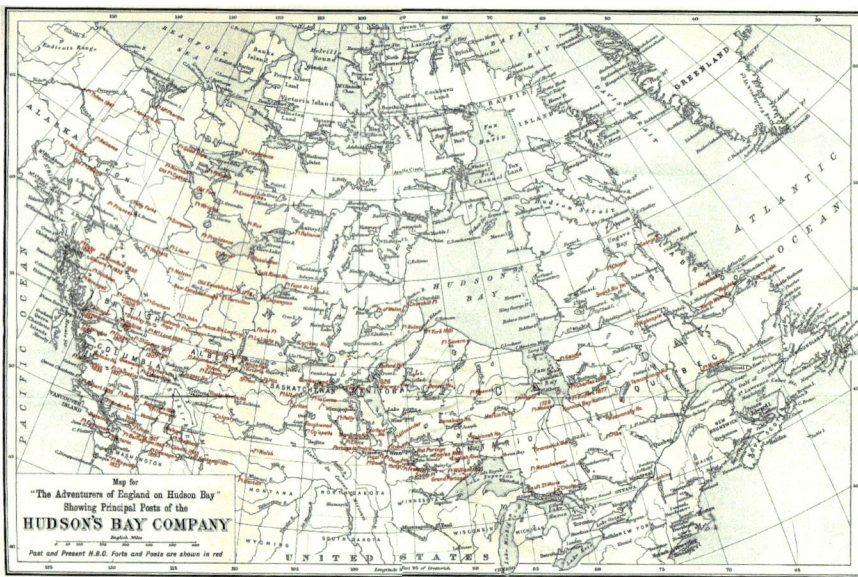

Figure 4.3 Hudson's Bay Company Map (Wikimedia Commons)

Hudson's Bay companies, serving as representatives of Great Britain, were successful for a prolonged period of time. These commercial entities filled a role that governmental agencies could not or would not assume on their own. However, the importance of these joint-stock companies to the home nation was much greater than simply their financial or even their political impact; they also served as exploring arms for Great Britain, strategically placed to encourage the medical explorers to venture forth from their centers of trade to investigate the world. Those companies' place in history is a significant component of this story, as throughout their existence they provided travel and employment opportunities for thousands of Scottish men. For dozens of medical explorers, a voyage to a company's commercial base of operation was often their earliest experience of foreign travel.

5

Joseph Banks: Send them forth

The works which Joseph Banks leaves behind him occupy a few pages only;
their importance is not greatly superior to their extent;
and yet his name will shine out with lustre
in the history of the sciences.

–Baron Georges Cuvier

Organizations such as the British Army, the Royal Navy, the East India Company, and the Hudson's Bay Company each played a critical role in what could be described as channels for the medical explorers. That is, these vital entities linked an employment opportunity in a foreign land with a platform that could launch these Edinburgh-trained doctors onto their various exploring missions. But on some occasions the connection was not an organization but an individual, and if one man could be labeled the father of the medical explorers it would be Sir Joseph Banks. **(Figure 5.1)**

In *Sightseers and Scholars*, Stephen Brown identifies four types of naturalists: Early Naturalists and Scientists, Inquisitive Aristocrats, Professional Collectors of the Golden Age, and The Last Field Naturalists. He states:

> In the early nineteenth century the study of natural history was primarily the province of the wealthy and aristocratic. With little government support for expeditions to distant lands, and with public interest in natural history just starting to expand … only wealthy individuals could organize extensive field excursions in pursuit of scientific specimens.[147]

Joseph Banks would certainly fit Brown's description of the Inquisitive Aristocrat. Banks was the son of a rich Lincolnshire country squire who was a member of the House of Commons. As a boy Joseph enjoyed exploring the countryside, and developed a keen interest in nature, history, and botany. A child of privilege, he was educated first at Harrow School and Eton College, then enrolled at the University of Oxford. When his father died in

1761, he inherited the family estate of Revesby Abbey, becoming the local squire. At that point he left Oxford without taking a degree.

Dividing his time between Lincolnshire and London, Banks continued his interest in science by working with the curator of the Chelsea Physic Garden and with Daniel Solander at the British Museum. Through these connections he began to make friends throughout the scientific community. His influence continued to increase, and he became an advisor to King George III, urging the monarch to support voyages of discovery. In 1766, Banks was honored to be elected to the Royal Society, the oldest national scientific institution in the world, and then was selected to participate in an exploring voyage to Newfoundland and Labrador. His reputation grew when he published the first descriptions of plants and

Figure 5.1 Sir Joseph Banks (Wikimedia Commons/PD)

animals collected on this voyage and documented 34 species of birds found only in these two regions.[148] His second exploring voyage was in 1768, when he was appointed to a joint Royal Navy/Royal Society expedition to the South Pacific aboard HMS *Endeavour*. This was the first of Captain James Cook's voyages in search of the *Terra Incognita Australis*—the mythical continent lurking somewhere in the southern reaches of the globe.[149] As a member of the Royal Society and an advisor to the king, and having already participated in one successful exploring voyage, Banks was a reasonable choice as the naturalist for this epic voyage. However, his social and political position, coupled with the fact that he personally funded eight others (a Swedish naturalist, a Finnish naturalist, two artists, and four servants from his estate) to join the expedition would undoubtedly have also played a role in his selection.

In 1771, following his participation in the well-publicized three-year voyage, Banks returned to England and instant fame.[150] His reputation now well established, he next focused his energy on two endeavors, which—unbeknown to him at the time—would launch the majority of explorers throughout the British Imperial Century. The first of these took place at the newly established Royal Botanic Gardens.

Royal Botanic Gardens Kew

The increase in maritime trade in the 16th and 17th centuries saw a huge surge in the number and variety of medicinal plants being brought back to Europe as trophies from distant lands. During this period flora began to be imported from North America, South America, and Asia, as well as from some of the more remote parts of Europe. The Royal Botanic Gardens Kew, founded in 1759, was initially a physic garden located on part of a royal estate. One of Kew's initial purposes was to provide medical students and practicing

Figure 5.2 Willliam Aiton (Wikimedia Commons/PD)

physicians with a ready supply of plants, many grown from the imported collections gathered by sailors returning to England from across the empire. At the time the practice of medicine required that the physician create his own medications, so a critical component of a doctor's education included knowledge related to the plant-based treatments which could be made from newly acquired specimens arriving from around the world.[151]

By the middle of the 18th century physic gardens were already moving away from their medicinal function in favor of other scientific and aesthetic priorities. William Aiton, a Scot from Hamilton,* was appointed Kew's first director, and is credited with changing the facility from a source for developing medicinal plants into an institution which came to be more strongly associated with research conducted in the herbarium, library, and laboratories housed within the gardens. **(Figure 5.2)**

Ships of the Royal Navy and merchant trading ships returned from their journeys with animals, plants, and natural curiosities that had never before been seen by the people of Great Britain. The collections, many of which were housed at Kew, provided the British scientific community with new and captivating research opportunities. But the rare and unique finds also expanded the knowledge base of the British public, who were increasingly eager to learn more about the exotic lands that now comprised the vast empire.

For much of the 18th and 19th centuries, Britons of all social classes became avid, although mostly amateur, natural historians. Their enthusiasm for natural history was abiding and passionate. Caged birds, exotic plants in conservatories or Wardian cases,† sea anemones in aquariums, and minerals in cabinets proudly displayed in private homes testified to the popularity of natural history.[152] Lynn Barber states:

> By the middle of the century there was hardly a middle-class drawing room in the country that did not contain an aquarium, a fern-case, a butterfly cabinet, a seaweed album, a shell collection, or some other evidence of a taste for natural history.[153] **(Figure 5.3)**

---

* Now a suburb of Glasgow

† A Wardian case was an early type of terrarium.

The interest in natural history was also reflected in an increased popularity of the museums, zoos, aquariums, and botanic gardens which were soon constructed in most major cities. These institutions organized outings to collect local species of plants, animals, and insects, and "botanizing" became a common pastime for Victorian Britain.

Figure 5.3 Wardian Case (Wikimedia Commons/PD)

While the amateur gatherers filled their drawing rooms with collections assembled from across the United Kingdom, a cohort of professional naturalists also appeared. These explorers literally traveled around the world in pursuit of previously unknown species of plants. The study of natural history was considered a science, seeking knowledge of the world around them. Naturalists were considered part of the scientific community, and the new plants and animals they collected were vital to a growing understanding of the natural world. However, no universities offered a degree in natural history.[154] A few universities had professors of natural science, but the discipline was typically scorned by the academics of the time. A Cambridge professor in the 1840s said that natural history "should be discouraged as much as possible and should be regarded as idle trifling." [155]

Eventually, different disciplines would develop. While natural science as such remained the observation and study of the natural environment, various distinct sciences such as botany, geology, biology, and zoology would emerge.[156] However until these disciplines started to appear in 1893, when the Faculty of Science was introduced at the University of Edinburgh, the Scottish universities elected to ignore the Cambridge professor's opinion; the best option for students interested in the natural sciences was medicine.[157] This was because the courses required for a degree in medicine exposed them to a wide variety of natural history topics that could not be found in any other university course of study. Thus, it is not surprising that with this exposure to the natural sciences, accompanied by the discipline's huge popularity throughout Great Britain, these doctors were attracted to a life where collecting, cataloguing, and analyzing the natural world would be their main charge. Although the doctors of Edinburgh were involved in a wide variety of exploratory endeavors, the vast majority of them were focused on botanical collecting. Ann Lindsay, in *Seeds of Blood and Beauty: Scottish Plant Explorers*, states:

> A handful of Scots … have hunted out and introduced into the West more plants from around the world than probably all the other European nations combined.[158]

Economic botany

William Aiton was greatly assisted in transforming Kew by King George III, who pursued agricultural improvement with great vigor. George, who had ascended the throne in 1760,

Figure 5.4 Palm House Royal Botanic Gardens, Kew (Wikimedia Commons/PD)

utilized Kew as both his experimental garden and as his part-time home; he lived at Kew for about three months each year, from the middle of May. He bred sheep there, and experimented with new crops that could enrich and feed his people, earning himself the nickname Farmer George.[159] He greatly enhanced and expanded Kew's original holdings by merging the royal estates of Richmond and Kew in 1772, and from that point forward the king's and Aiton's emphasis within the garden shifted to what became known as "economic botany." **(Figure 5.4)**

Economic botany is the study of the relationship between people and plants; it explores the ways humans can use plants for food, medicine, and commerce.[160] While the general public reveled in viewing the exotic species now regularly on display within the gardens, King George understood the extent to which the empire was built on plants of economic importance (tobacco, cotton, and spices among others). The importance of plants in expanding and developing the empire is reflected in a quote from the *Pall Mall Gazette* in 1891:

> A competent and capable botanist may do more to open up the country than a dozen mining engineers, for the discovery of a single plant useful to commerce may be of greater value to Africa than many gold mines.[161]

Banks continued to be a scientific consultant for King George III, and shortly after his return from the South Pacific he began serving as the "unofficial advisor at Kew Gardens." Reflecting the tenet of economic botany, Banks and Aiton were soon dispatching plant-collecting explorers to the Southern African Cape, Australia, Chile, China, Ceylon,* Brazil, and elsewhere. But as their access to both personnel and ships was limited their efforts were haphazard. Banks solved this problem by turning to his friend John Montague, First Lord of the Admiralty. Once Banks, through his urging, had convinced Montague to begin placing botanists on board each of His Majesty's ships, no longer would plants arrive by chance after being picked up from wherever a ship had happened to put in. Now organized expeditions were taking place, with specialist gardeners, or surgeons with botanical training, aboard. Royal Navy ships dispatched to specific corners of the globe were directed to bring back useful and valuable plants to Kew.[162] Banks' position at the Royal Botanic Gardens Kew was formalized in 1797; under his leadership the British Empire continued to enjoy a golden age of plant exploration, and the Royal Botanic Gardens Kew became "the great botanical exchange house of the British Empire." Botanic gardens within the British Isles, and those

---

* Now Sri Lanka

established throughout the British Empire, became major sites of scientific research, and many species were introduced to Europe after first passing through Kew Gardens.[163] But Banks was visualizing even bigger dreams.

Hub and spokes

By the end of the 18th century Great Britain was already well ahead of the rest of the world in plant collecting, and throughout the 19th century it remained the largest purveyor of plants. But while the accumulated specimens greatly expanded the scientific mission of the garden, the impact of plant discoveries on economic botany was inconsistent. When the plants were being collected in climates similar to that of Great Britain they could be grown in England on a large scale and the commercial value was immediate. But many plants discovered in the empire's tropical climates could, when carried back to England, only survive in greenhouses. The small numbers of plants that could be grown indoors did not make economic botany cost-effective, so King George, in collaboration with Sir Joseph Banks, who was now his official advisor at Kew, began to look at a different method of plant distribution.

Banks made no secret of his belief that exploration preceded colonization, and his contemporary Robert Hay, Permanent Under-Secretary at the Colonial Office, admiringly declared that Banks "was the staunchest imperialist of the day." [164] Banks envisioned a network of botanic gardens around the empire with Kew at its center and plants passing through it back and forth between the gardens: he called this the "hub and spokes" structure. Through his power and influence at Kew he was able to dispatch medical explorers to be relocated permanently at remote areas of the expanding empire. There they could continue their passion for botany and natural science by establishing botanical gardens that would serve as the spokes to Kew's hub. The late 18th and early 19th centuries were marked by the establishment of numerous botanical gardens in territories recently acquired by Great Britain and now under its control, the majority of them located in lands under the management of the British East India Company.

Banks, whose position at Kew had moved from advisory to almost total control, saw these gardens as essential components in securing both profit and power for the empire through the global transfer of plants. As he explained to the superintendent of the Botanic Garden at St. Vincent:

> (They) might in the future be of great utility to the public both by the improvement of many valuable plants, growing wild and neglected in the British colonies, and by the introduction of many articles of value in a commercial or medicinal view, only produced in foreign settlements, and not to be procured by the British but at very high prices, especially in times of war.[165]

Medical doctors, most of whom had earned their degrees in Edinburgh, were now routinely being appointed as superintendents of botanic gardens and to other senior posts within the colonies. The sheer number of Scottish doctors in such posts from the late 18th to the late 19th century acknowledges the value that the empire placed on these uniquely trained individuals. Other countries were also establishing botanic centers within their colonies; in addition to the English there were French, German, Dutch, and Swedish botanists. However, in the Victorian era 80 percent of all colonial botanic gardens in the world were British, and in terms of sheer numbers the Scots, and the Scottish-trained botanist-physicians, dominated the management of these facilities.[166]

Joseph Banks' imperial leanings and abiding interest in botany coalesced in a resolve to transform the Royal Botanic Gardens Kew and the various colonial satellite gardens "into a great botanical exchange house for the empire." These early regional centers, and the many that followed, were instrumental in developing the trade in cloves, cinchona,[*] sugar, palm oil, coffee, teak, tea, breadfruit, pepper, and rubber. Now, although many seedlings were sent from the colonial botanic centers to the Royal Botanic Gardens Kew, the directional flow of plants was no longer exclusively from the spokes to the hub. Nor were the products in the spokes developed exclusively for local use or for export directly to England. During this period breadfruit was transferred from the South Pacific to St. Vincent in the West Indies; tea from China to India and Africa; rubber from South America to Malaya; mahogany from the West Indies to India; cocoa from the West Indies to Africa; the oil palm from Africa to Malaya; cinchona from South America to India; and the American pineapple to all the tropical regions of the empire. These types of plant transfer continued into the middle of the 20th century and had huge economic consequences around the world.[167] The coordination of this new approach—that is, direct from spoke to spoke—became an additional function of the medical explorers in charge of the scattered botanic gardens.

Botanical expansion

During the late 18th century and throughout the 19th, Great Britain used the establishment of tropical botanical gardens as a tool of colonial expansion. The primary focus of these gardens was trade and commerce, but they soon expanded to the secondary focus of science. The hub and spokes system created by Banks was now securely in place, and the Royal Navy's medical explorers were firmly ensconced in every aspect of economic botany. The final step that would make the Royal Botanic Gardens Kew the greatest botanic center in the world was about to be put into place.

Kew had traditionally received a share of the plants collected by sailors and transported aboard Royal Navy ships, and from 1840 this was the unofficial policy enacted by the Admiralty. Then from 1854 it was a governmental requirement that Kew become the exclusive destination for all collections made at government expense. Lucile Brockway, in *Science and Colonial Expansion: The Role of the British Royal Botanic Gardens*, states that the government's decision to place Kew ahead of the other botanic gardens solidified its position as the world's center of botanical research:

---

[*] quinine

These exploratory voyages of the 18th and 19th centuries, to which naturalists were attached … constitute an almost unrecognized government subsidy to natural science as important as any more formal government subsidy to science in our own time.[168]

Banks found both Kew and the colonial botanic gardens to be useful centers for the acquisition and propagation of economic crops.[169] He viewed the gardens that Kew controlled through his hub and spokes system both as instruments of imperial endeavor and as a legitimate means of developing the natural resources of the empire. These gardens were established as a result of two major factors. First, imperialistic expansion was fueled by economic botany. Second, the shift in exploration from trade to the search for scientific knowledge for its own sake emerged in the wake of the Scottish Enlightenment.[170] In the century or so during which the physician explorers flourished, botany became a respected professional career rather than the domain of gardeners and amateurs. Plant collecting and cataloguing had reached their zenith, and Kew, located at the center of an international network of co-operating gardens, became arguably the most prestigious botanic garden in the world. The intercontinental transfer of economically important plants transformed the world economy forever. Kew was now not only the hub of economic botany, but it was at the center of the British Empire. Through the efforts of Joseph Banks, it had become first among equals.[171] But as impactful as Joseph Banks' efforts were to Kew, his second endeavor would play an even larger role in launching the medical explorers.

The Saturday's Club

In 1788, a dozen members of the upper class met for an evening of dining and discussion at a tavern known as the St. Albans, near the eastern end of Pall Mall. Their leader was Sir Joseph Banks, and they called themselves the Saturday's Club. The majority of the diners that evening were members of the Royal Society of London for the Improvement of Natural Knowledge. Begun in 1665 by a group of physicians and philosophers, the Royal Society, as it became known, had been created for the improvement of scientific knowledge. Over the decades since its founding, its members included some of the most noted scientists of the day, including Sir Christopher Wren, Sir Isaac Newton, and Robert Hooke. In 1778 Joseph Banks became president of the Royal Society, a position he would hold for the next 42 years.

The members of the Royal Society, by encouraging what they termed "scientific geography," had been urging the government to support expeditions of discovery. In doing so they also promoted special projects related to the creation of navigational charts and maps. As a result, the Royal Society began to be more greatly involved, both at home and abroad, in the growth and development of geographic endeavors. Banks, however, felt that the organization was far too occupied with the advancement of other branches of science, and that geography was not receiving enough attention. As president, he felt there was an urgent need for a central organization that would guide, control, and advance the study and business of exploration.[172] That evening in the tavern he had selected dinner companions who shared his passion for geographical study. It was he who focused the conversation on the need for concentrating British exploration on what the members of the Saturday's Club felt was the greatest failing of their time; men were sailing around the world, yet the geography of the interior of Africa remained almost entirely uncharted.

A club member reflected:

> That as no species of information is more ardently desired, or more generally useful, than that which improves the science of geography; and as the vast Continent of Africa is still to a great measure unexplored, we should form ourselves into an association for promoting the discovery of that quarter of the world.

Henry Beaufoy, the club secretary, wrote, "Almost the whole of Africa is unvisited and unknown," and he added that a map of its interior was "still but a wide blank." On that evening, the Saturday's Club became the Association for Promoting the Discovery of the Interior Parts of Africa, which then became known commonly as the African Association. Banks was elected its secretary. Its meetings were often held in his home, and he would maintain the position of secretary for nine years, until 1797, when his other pursuits, especially those involving Kew Gardens, obliged him to resign from his duties in the association. However, he then assumed the role of treasurer, a position he held until his death.

The formation of this group was effectively the "beginning of the age of African exploration and this group of wealthy businessmen, possessing a keen interest in learning the unknown, would ultimately prove successful." [173] However, penetration of the Dark Continent would not happen quickly.

Exploration for the sake of geographic knowledge was the group's primary concern. However, some of the titans of British industry were members, and to them securing a trade advantage for themselves and for England was part of the plan. "Gold is there so plentiful," a member of the Association's Committee wrote at their second meeting, "as to adorn the slaves … If we could get our manufactures into that country, we should soon have gold enough." These entrepreneurs also saw the reduction or elimination of slavery as linked to their success in business. According to Davidson in *The African Past: Chronicles from Antiquity to Modern Times*, the two great questions to be resolved at the end of the 18th century were how England could further reduce the overseas slave trade by attacking it at its source, and the discovery of trading opportunities beyond the coastal barriers that had been established by the middleman chiefs.[174]

Although the African Association was motivated by a quest for knowledge and a romantic love for adventure, its members were also inspired by the rising value of African markets and access to the raw materials produced by the continent. It was also a common belief of the time that in order to eradicate slavery the African rulers would need to have the income provided by human trafficking replaced by a more socially and morally acceptable revenue stream. The African Association felt that by establishing legitimate trade with those inhabiting the area from which the enslaved people were being collected, the need for the rulers to "sell their fellow man" would be reduced or eliminated. At the same time, the profits of the British industrialists would soar.

The association's work would also be in the interest of Christian learning. In the same way that the profit from slavery needed to be replaced by commercial trade of equal value, introduction of Christian morals to a people practicing Islam or paganism was seen by the association's members as a vital tool for combating the trading of humans. "In the pursuit of these advantages," Beaufoy emphasized, benefits would at the same time be "imparted to nations hitherto consigned to hopelessness and uniform contempt." [175]

THE GRAND MOSQUE OF TIMBUCTOO

Figure 5.5 Timbuktu
(Wikimedia Commons/PD)

Thus, the African Association was formed by a group of individuals representing political, commercial, scientific, religious, and abolitionist causes. Each member pledged to contribute five guineas* per year to recruit explorers and fund expeditions from England to Africa. Although their interests were not necessarily compatible, they agreed that to a collective end they would concentrate on Africa in general and West Africa in particular, and it would be Sir Joseph Banks who provided the focus that would keep the African Association moving forward.

Timbuktu and the Niger River

To promote any of their aims, the African Association needed first to explore a continent that was generally unknown to the outside world. After collecting as much information as possible, they began recruiting explorers for West Africa using a two-pronged approach for the enterprise. Most of these brave individuals would embark from already well-established ports along the north and east sides of the continent; leaving from sites such as Cairo, Tripoli, or Zanzibar they would attempt to reach West Africa by utilizing the caravan routes established by Arab traders. Others would use the various rivers that emptied into the Atlantic along the western coast, attempting to find the way to the center of the continent by following those previously unexplored waterways.

Each of these approaches shared a common goal: to locate the fabled city of Timbuktu, the "lost city" of gold. **(Figure 5.5)** This legendary city was said to be located on a large river known to the Europeans as the Niger; finding the course and source of the river meant finding the city. It was believed that from Timbuktu flowed exports of gold in such quantities

---

* A prestigious form of British currency, worth £1 1s, ie 105 percent of £1 sterling. While guinea coins—solid gold—are now collectors' items, the term is still used to refer to the prize money in horse racing and the prices in high-end auctions.

that the city took on the reputation within the outside world of possessing endless wealth.[176] To Europeans fascinated by the discovery of new worlds, Timbuktu's unimaginable and scintillating trade opportunities were too great a temptation to resist. Peter Brent, writing in *Black Nile: Mungo Park and the Search for the Niger*, asserts:

> The state that controlled the Niger traffic controlled the flow of trade; with the western Sahara route disused, shipments loaded or unloaded at Timbuktu could be carried along the central and eastern desert routes connecting the Niger with the Mediterranean countries. Domination of the Niger was worth fighting for.[177]

In the 18th century no European had ever seen the river itself, and in fact many were convinced it did not exist. Yet it was well known to the non-European world and had been traveled by Arab traders for hundreds of years. It had long been the major highway of commerce between the kingdoms of Africa's interior and traders from Arab nations as far away as present-day Iraq. But the maps available to the African Association members showing its origin, directional flow and mouth were mere guesses based on second- or third-hand stories told by Arab traders—who had obvious reasons for discouraging European exploration of the continent's interior.

With a lack of accurate charts and maps, a wide variety of conflicting stories emerged regarding both Timbuktu and its fabled river. Davidson Nicol, the Sierra Leone academic and diplomat, describes the following:

> The most popular description from the 16th to 18th century was one in which the river rose from a lake near the Equator in the centre of Africa, the Lacus Niger. From this point it was supposed to flow northwards almost in a straight line to reach another large lake, the Lacus Bornu. Before reaching this, it was said to flow underground for a distance variously given as being between 18 and 60 miles. After Lake Bornu, it took a bend of 90 degrees and flowed westwards through another lake, Sigisma or Guarde, to break eventually after another lake system into four rivers, amongst which were the Senegal and the Gambia, which all emptied into the Atlantic at the western most point of Africa.[178]

Thus, the mission of the African Association evolved from the general exploration of the continent to a focus on West Africa and to a search dedicated to locating and exploring the Niger. All those involved felt certain that the discovery and exploration of the river would provide the Association's members with the keys to Timbuktu.

Trial and error

A study of the first explorers recruited by the African Association reveals that they were generally ill-suited and underprepared for their task. It is also clear that most of these individuals either swiftly gave up and returned to England or made valiant efforts that quickly led to their deaths. Whether attempting to follow the Arab trade routes or attempting to follow the course of an African river, the one consistency these early explorers confronted was the brutally hostile environment of Africa: its adverse climate, difficult terrain, and prevalence of numerous fatal diseases. Most of these early explorers died of fever, exposure, or fatigue. But in addition, the hostile environment extended to the people they met. These were the people the explorers needed to rely upon for food, shelter, direction, and safe passage—and many of the early travelers died at the hands of those local people.

The African Association's initial explorers found it difficult, if not impossible, to establish trust with the people they met in Africa. To us now, this is unsurprising, because from the African perspective, how could those travelers be trusted? The coastal Africans suspected them of trying to take away their livelihood by initiating direct trade with the people of the interior. Then the Arab slavers, still collecting as many enslaved people as ever from the people of the interior, correctly assumed that a primary objective of these travelers was to abolish slavery altogether. Yet these men from England were saying that their only reason for penetrating the continent was to seek knowledge. Both the Africans and Arabs thought it was another example of the white man's cunning. It was this atmosphere of fear, competition, and jealousy that added immensely to the difficult task of the first explorers, and cost many of them their lives.[179]

Nevertheless, the African Association wasted no time, and in the first year of the group's existence the first explorer, an American named John Ledyard, was recruited for travel to Africa. He had traveled around the world with Captain Cook and had also traveled extensively in both Russia and North America. With his previous travel and exploration credentials he appeared to be an ideal choice for the association, and he sailed from England to Cairo, arriving in 1788. His adventure was short-lived. While preparing for his overland journey he became ill. Rather than seeking medical attention, he self-medicated, poisoned himself with a fatal dose of sulfuric acid, and died in Cairo.

Perhaps anticipating the worst, before Ledyard had even arrived in Cairo the African Association had enlisted one Simon Lucas to attempt a mission, starting from Tripoli. Lucas spoke fluent Arabic, had spent time in Morocco and was already acquainted with the British ambassador in Tripoli. Lucas found guides to take him across the Libyan Desert, but soon after his departure his guides abandoned him, and he was forced to make his own way back to England. The African Association next sent an Irishman, Major Daniel Houghton, to proceed from the mouth of the River Gambia inland towards what was hoped to be the Niger River. He penetrated farther into Africa than any European before him and reached the highest navigable point on the Gambia. He then continued on foot toward Bundu. Unknown to him, he managed to get within 160 miles of the Niger and 500 miles from Timbuktu, but in September 1791 he was lured into the desert by his local guides, robbed, and killed.[180]

Those three attempts to locate the Niger River had resulted in two explorers dying and the third abandoned in the desert. The African Association would finally succeed with the recruitment of its first medical explorer.

Mungo Park was an Edinburgh-trained doctor recruited directly by Joseph Banks on behalf of the Association. Park had made a successful exploring and collecting voyage while in the employ of the East India Company. His skill as an explorer and naturalist had impressed Banks, and following his return from an East India voyage to Sumatra, he was employed by the Association. He left England on the trade ship *Endeavour* and arrived on the West African coast in June 1795. He followed the route along the River Gambia established by Houghton, and after surviving several near-fatal encounters with the local inhabitants was finally guided to Segu (in present-day Mali) by some friendly Bambara people. It was at this location that Mungo Park became the first European to see the fabled Niger River, and to record that at that point it flowed toward the east. His plan was to follow the river to Timbuktu, and he followed its course on foot for over 75 miles. But he was not to progress further. He was prevented from advancing along the river when he encountered a new set of hostile locals, who robbed him of his supplies and remaining trade goods. So, he was forced to return to England—but he returned as an instant national hero, and news of his success caused membership within the African Association to swell dramatically.

Mungo Park's travels and discoveries had the greatest impact upon Western knowledge of the African continent to that point. Frank Kryza writes:

> News of Park's accomplishments thrilled the African Association and indeed all of England. He was the first white man to penetrate the forbidding interior of Africa for the sole purpose of finding out what lay there, and to come back alive. He invented a new and glorious calling, creating an adventurous species of hero: the lone, brave African explorer: the African traveler. This ideal soon captured the imagination, fed the fantasies and fulfilled the literature of Europe.[181]

Park's mission to Africa had lasted two and a half years, and he published his account of the undertaking in 1799; the work was hugely popular and remains in print today. He attempted to return to a typical physician's life by marrying and establishing a medical practice in his native Scotland. But the comfort and security of such a life was not to last. Restless, he asked the African Association to assign him to another exploring voyage. They agreed, and Park attempted a second expedition to reach Timbuktu in 1805. This required the employment of a much larger group, comprised primarily of military personnel already stationed in West Africa. Park followed his earlier route and once again reached the Niger. But by that point his initial large body of men, suffering almost constantly from malarial fever, had been reduced to less than ten.

The remaining members of the party began building a boat in which they planned to travel with the current until they reached the river's terminus. But more men died, and the group which finally set sail downstream had shrunk to just four men. Within a few days of his departure, Park successfully sailed past Timbuktu and ascertained that just beyond the fabled city the river made a large bend, then headed south-southeast. After traveling on the river for nearly 1,500 miles his party was ambushed near Bussa, in present-day northern Nigeria, and except for their African guide, Amadi Fatouma, everyone was killed.

So, Park did not live to see the termination point where the Niger empties into the Gulf of Guinea, part of the Atlantic Ocean. But Amadi reported to others that Park had determined the ultimate directional flow of the river, to the south-southeast. In addition, from the safety of his boat, Park had viewed the "golden city" itself. But as he had died before he was able to share his discoveries with the world, everything he had seen and documented remained "undiscovered."

The Royal Geographical Society

Following Park's second expedition African exploration was drastically reduced for over a decade. With the Napoleonic Wars (1803–1815) Britain was too preoccupied with France to plan and implement exploring voyages. However, with the end of the conflict the British government elected to take on a larger role in order to establish commercial domination in Africa.

By this time Sir Joseph Banks was growing ill. Although mentally as acute as ever, from about 1805 on he lost the use of his legs and had to be wheeled to and from his various meetings. His health continued to decline, and in 1820, six months after King George III had died, Joseph Banks passed away at his home in London. He was 77 years old.

By the time of his passing, Banks had cemented his reputation as one dedicated to the study of geography and natural science. His wealth and position had enabled him, as one of the Inquisitive Aristocrats, to travel around the world collecting specimens. His work with King George III, along with his work and collaboration with the Royal Botanic Gardens Kew, had led to the development of the extensive and highly successful hub and spokes system. His 42-year tenure as the President of the Royal Society had placed him in the vanguard of scientific exploration, and his love for the adventure of discovery had led to the founding of the African Association. Joseph Banks was not a scientist. He was a self-educated generalist. But his efforts arguably did more than those of any individual to launch the era of the surgeon explorers.

Following his death the Association continued, but its influence began to diminish. One of its last major accomplishments involved Alexander Gordon Laing, a major in the Royal African Corps. He believed he had found the source of the Niger and proposed traveling along its course to locate its mouth. The African Association supported his project, and in February 1825 Laing left England for Tripoli. Taking months to cross the Sahara, suffering from fevers and the plundering of his caravan, he managed to reach Sidi al Mukhtar's settlement at Azawad, about 250 miles from Timbuktu. There he joined another caravan, and in September 1826 became the first European to walk through the gates of Timbuktu. After a brief rest he began his return journey. Two days after leaving Timbuktu he was murdered by his Arab guides.

For the near-50 years of its existence the African Association did much to promote the exploration of West Africa. But then in 1830, under the patronage of King William IV, the Geographical Society of London was founded as an institution to promote the "advancement of geographical science." [182] The original purpose of the Geographical Society was to collect, digest, and publish interesting and useful geographical facts and discoveries; to accumulate a collection of books on geography, voyages, and travels, and of maps and charts; to keep specimens of such instruments as are most serviceable to a traveller; to afford assistance, instructions, and advice to explorers; and to correspond with other bodies or individuals engaged in geographical pursuits.[183]

Rather than having a focus on a single region, the aim of the Geographical Society was global exploration, and in 1831, when the society absorbed the African Association within its membership, the organization that had begun life as the Saturday's Club ceased to exist. In 1859 Queen Victoria granted the Geographical Society a Royal Charter, so it became the Royal Geographical Society. It was this prestigious organization, which exists to this day, that sponsored almost all of the exploration efforts during the British Imperial Century.

6

Archibald Menzies (1754–1842)

Nature is an inexhaustible source of investigation … ,
she presents herself to those who know how to
interrogate her, under forms which they have never yet examined.

–Alexander von Humboldt

The reasons for exploring during the British Imperial Century were as varied as the interests that had led to the formation of the African Association. The enhancement of commerce, the spread of Christianity, the eradication of slavery, and the advancement of geographic knowledge and scientific development were all reasons for the medical explorers to venture forth. However, by far the largest number of these intrepid Scottish travelers were naturalists, identifying and collecting plants from around the world that had been previously unknown to Europeans.

Plant collecting was a risky and hazardous business. Disease, accidents, and attacks by the local populations resulted in many naturalists meeting an early death. As Carl Linnaeus wrote in *Glory of the Scientist*:

> Good God, when I consider the melancholy fate of so many of botany's votaries, I am tempted to ask whether men are in their right mind who so desperately risk life and everything else through the love of collecting plants.

The question of why those men risked their lives in their quest for new, unusual, or perhaps valuable plants is difficult to answer. While they might gain personal credit from other naturalists, very few of them would achieve fame, let alone fortune. Although the motivations remain obscure, there appears among almost all these collectors a genuine curiosity and deep personal interest in their subject. This was often combined with a sense of adventure and a desire for exploring unknown territories.[184]

These individuals shared much in common: most were Scotsmen, educated in a Scottish university medical program, who had developed an interest in natural science at an early age.

But they did not come from any particular geographic region of Scotland, so the significant influence seems to have been national rather than regional. There was never a cluster of naturalists who came from a particular area of Scotland in the way that Orkney produced so many of the medical explorers associated with the Hudson's Bay Company.[185] Rather, the plant collectors rose through a national, and even international, interest in the natural sciences which had already been established well ahead of the British Imperial Century. It was said that if one was looking for the top gardener for an estate, botanic garden, or exploring voyage—find a Scotsman.

Archibald Menzies,[*] born in 1754 in Weem, in the parish of Newhall, was among a large group of medical explorers who made their mark as naturalists and what Stephen Brown would call one of the "professional collectors." Archibald's parents, James and Anne, had five sons and three daughters. The timing and location of Archibald's birth were fortunate, as Weem and the surrounding area, including Newhall, were at that time quite prosperous. Menzies and his large family lived on the family croft, a small rented farm consisting of a house in a plot of land, with a right to additional pasturage held in common with other surrounding farms. In this case, each of the numerous crofts surrounding Newhall were the property of the local laird, Sir Robert Menzies.

It was not uncommon in Scotland to have members of the same extended family living in close proximity. Direct descendants of the head of the clan inherited title and land, but there was also an attempt to see that even the cadet[†] members of the family were well provided for. A search of the records shows that during Menzies' father's time, half of the staff employed by Sir Robert bore the surname Menzies. Although Archibald, his father, and his brothers tended the land, the family was not dependent on the croft for their livelihood, as James was also a gardener at Castle Menzies, the home of the laird. Gardening was a strong tradition in the Menzies family, and all of Archibald's brothers—William, Robert, John, and James—worked as gardeners throughout their lives. When they grew older, both Archibald and an older brother, William, were employed on the grounds of Castle Menzies, working with their father.[186] So, it was at Castle Menzies that Archibald developed both his interest and skills in gardening.

John Hope and the Royal Botanic Garden Edinburgh

When Menzies was aged about 20, his employment moved from Castle Menzies to the Royal Botanic Garden Edinburgh (RBGE). From 1774 to 1778 his was a familiar face around the garden. His brother William had begun working at the facility several years earlier, and the garden managers' familiarity with William and an appreciation of his work undoubtedly made a position for Archibald much easier to secure. His application for employment as an assistant gardener undoubtedly included a recommendation from Sir Robert, which would also have strengthened his submission. Then, whilst he was working at the botanic garden.

---

* Pronounced *ming*-iss
† Junior

his burgeoning gardening career had an unexpected boost; both Archibald and William came under the considerable influence of Dr. John Hope, the Professor of Botany at the University of Edinburgh. (**Figure 6.1**)

Hope was nearing the end of his career. Although his motives have never been clarified, perhaps he saw promise in Archibald, for during the young man's time at RBGE the professor repeatedly interceded on his behalf. He first invited him to sit in on his botany lectures. Professors at that time were paid a small salary supplemented by collecting fees from each student who attended the lectures. It is clear that Menzies would not have been able to pay these fees out of his meager salary as an assistant gardener. But university records make it apparent that he participated in several of the professor's courses. Possibly John Hope waived the fees, or conceivably the Royal Botanic Garden Edinburgh

Figure 6.1 John Hope (Wikimedia Commons/PD)

might have seen value in Menzies pursuing the study of botany and supported the effort by sharing the costs. Although the way Menzies paid for these courses, and why others took such an interest in him at this early age, remain unclear, what is certain is that Dr. Hope also encouraged Menzies to begin studying medicine at that time.[187]

Menzies did not complete his medical degree while studying at Edinburgh. However, he had clearly proved himself an able student, and by 1781 he had been granted a license to serve as an assistant naval surgeon. He did not enter the military at once. He first traveled to Caernarfon in Wales, where he assisted a surgeon in an established practice. Perhaps, though, he found life in Wales a bit too dull, for not long after his arrival he took the route of many a Scot and applied for a position in the Royal Navy.

At that time Great Britain was still deeply involved in the American Revolutionary War, and America's greatest ally in this conflict was England's long-standing enemy, France. Menzies was initially appointed assistant surgeon onboard HMS *Nonsuch*, part of a British fleet that sailed for the West Indies to protect the sugar islands from a takeover by the French.[188] Good fortune had presented itself thus far in Menzies' short life. But that good fortune was about to change. Although the timing and location of his birth and his relocation to Edinburgh might have been fortunate, Menzies was not as lucky with the timing of his entrance into the Royal Navy.

Trial by combat

Ships' surgeons were responsible for all aspects of the men's health. Menzies' initial duties would have included inspections of all new recruits both for their fitness and for any indication of infectious disease. If the ship was engaged in battle he would have served in the capacity of a surgeon, but for the majority of time he would tend the men as a general practitioner.

Traveling from England to North America would have involved almost constant attention to the treatment of infections and scurvy, and despite Menzies' best efforts, 50 of his shipmates died before the ship reached the West Indies. He would also have discovered that the difference in social status between physicians and surgeons in Edinburgh extended to those serving in the Royal Navy. Within the ship's crew, those holding the rank of midshipman and above were regarded as gentlemen and were treated accordingly. But surgeons, and to an even greater extent assistant surgeons, were thought of as craftsmen who had certain manual skills. Although much better paid than most of the ship's officers, their status was seen as similar to that of the ship's carpenter, so they were not held in particularly high regard. While other officers might have all of their physical needs accommodated, surgeons were left to fend for themselves. Adding to the insult, they were expected to provide their instruments and medicines at their own expense. Menzies would have realized that he was not to be numbered among the gentlemen—and he would have encountered all of this before the *Nonsuch* reached the war zone.[189]

The American Revolutionary War was in its final phase and the French navy had provided the primary support for the rebelling colonies. François Joseph Paul, Comte de Grasse, had commanded the French fleet at the Battle of the Chesapeake in the autumn of 1781, and his victory there had led directly to the British surrender at Yorktown, assuring the rebels' victory overall. So now, the British, realizing they had lost control of the Thirteen Colonies, were concentrating their efforts on protecting their territory in the West Indies, not wanting to lose those colonies to the French while the United States was gaining its independence.

After his victory at Chesapeake, de Grasse moved his fleet to the Caribbean and was soon confronted by the newly arriving British fleet, including the *Nonsuch*. So, Menzies, from the moment of his arrival in the West Indies, was involved in a nearly continuous series of battles. He spent almost five months amputating limbs, caring for the severe wounds created by flying splinters of wood and metal, and watching as the steamy heat accelerated disease among the badly injured sailors. Throughout the British fleet during its defense of the West Indies, 266 men were killed and 810 wounded.[190] Finally, in 1782 the British decisively defeated de Grasse at the Battle of the Saintes near the island of Dominica, and for Menzies the nightmare of war ended.

In 1783 the Treaty of Paris ended the American Revolutionary War. The treaties dictated that the British would lose their Thirteen Colonies. This event marked the end of the First British Empire.

But the defeat of the French in the West Indies meant that the British would maintain control of their valuable sugar-producing colonies, and that the British fleet, with the end of hostilities, could disperse to other locations. Menzies was assigned to HMS *Assistance*, and the ship, after visiting many of the islands within the West Indies, began sailing north along the east coast of the United States. It then moved into the protection of British-held territory and docked at Halifax, Nova Scotia. For the first time since enlisting in the Royal Navy, Menzies found the time and opportunity to collect and document the unusual plants he found within the region.

John Hope and Joseph Banks

While Menzies' background in the natural sciences, along with his medical preparation, had allowed him to attain a position as an assistant surgeon in the Royal Navy it was only now that this son of a Scottish crofter was at last able to indulge in his true interest—collecting plants.

Up to this point, Menzies had owed much of his success to John Hope and the Royal Botanic Garden Edinburgh for providing him with funding and access to relevant studies. While his life so far had been successful, it seems he hoped to achieve more. He would do so by aligning himself with the Royal Botanic Gardens Kew and with Sir Joseph Banks, and he would once again turn to his mentor, Dr. Hope, to help achieve this goal.

Although Menzies' professional status, an assistant surgeon with little in the way of an established reputation as a naturalist, was low, he possessed three distinct advantages. First was his acknowledged success as a gardener at the RBGE. Second, his participation in courses in botany and his experience as an assistant surgeon placed him among the most qualified individuals from which the Royal Navy would be selecting men to serve as surgeon/naturalists on its exploring voyages. Third, and more importantly, Hope had known Menzies, and served as his mentor, for years. At that time, Hope was serving as Regius Keeper of the Royal Botanic Garden Edinburgh and also as the King's Botanist. In addition, he was Professor of Medicine and Botany within the University of Edinburgh Medical School. Perhaps most importantly, one of Hope's distinguished supporters was Joseph Banks. The strength of their relationship can be seen in Hope's earlier request for Banks' assistance in securing funds from the Royal Treasury in support of the RBGE; Banks had complied, and asked King George III to intercede. Thus, the garden had been fully funded.

Hope clearly had Banks' support and respect. On August 22, 1786, Hope wrote to Banks, personally recommending Menzies as a young naturalist with great abilities and even greater potential.[191] If John Hope had faith in Menzies, how could Banks fail to listen to what the young man had to say? Hope then notified Menzies, still in Nova Scotia, that he had recommended him, and Menzies began an extensive correspondence with Banks. In these letters he assured Banks that his primary interest in life was botany, and the natural sciences as opposed to the practice of medicine. He offered to search for specific plants that Banks might want for his collection. While he cultivated this new connection, he was careful not to sever his ties with Hope; Menzies collected samples of plants and seeds, which he prudently dispatched to both men.

He maintained his collection schedule and correspondence with Banks and Hope until he was faced with a dilemma. The fleet was being divided and Menzies was faced with three destination choices: he could remain in Canada, return to the Bahamas, or travel to England. As it was the dry season in Nova Scotia and the Bahamas, Menzies felt the prospects of gathering additional plants or seeds there would not be promising, and so he sailed for England.[192]

The *Prince of Wales* and the *Princess Royal*

Archibald Menzies had thoroughly enjoyed his time collecting while in Nova Scotia, but up to that point his experiences in this field had been limited. However, shortly after returning to England he became aware of an opportunity that would afford him both an extended period of collecting and an opportunity to travel to parts of the world he had only dreamed of visiting. He learned that as a result of Joseph Banks urging John Montague, First Lord of the Admiralty, to begin placing botanists on board each of His Majesty's expedition ships, the Royal Navy's edict had recently been issued. In 1786 Menzies would be among the first of the botanists to reap the benefit from this new regulation. In September the *Prince of Wales* and the *Princess Royal* were to sail from England down through the Atlantic, and round Cape Horn, bound for the Northwest Coast of North America, continuing to Hawaii and then to China. The two vessels would then, continuing westward, circumnavigate the globe. Menzies, recognizing the importance of this voyage and the impact it could have on his career, instantly contacted Joseph Banks:

> I am informed that there is a ship, a private adventurer now fitting out … to go around the world. Should I be so happy as to be appointed surgeon … [It would] afford one of the best opportunities of collecting seeds and other objects of natural history for you and the rest of my friends.[193]

Banks immediately provided a recommendation. It is unclear if the endorsement went to the Admiralty or to the director of the company sponsoring the expedition. What is clear, however, is that the reply was quick in arriving; Menzies was appointed surgeon and naturalist aboard the *Prince of Wales* under the command of Captain James Colnett, an experienced and capable officer who had already circumnavigated the globe. He had joined the Royal Navy in 1770, and as a midshipman on HMS *Scorpion* had come into contact with James Cook; in 1771 both Cook and Colnett moved to HMS *Resolution*, and Colnett participated in Cook's second voyage to the Pacific Ocean between 1772 and 1775. Colnett had continued to serve in the Royal Navy during the American Revolutionary War. But in 1786, with the end of hostilities, the work for naval officers was greatly reduced, and Colnett was forced to go on half-pay. So, that same year he agreed to command a two-vessel fur-trading expedition for the Etches Company. This was the voyage of the *Prince of Wales* and the *Princess Royal*, for which Menzies had been chosen.[194]

Because it was a privately funded expedition, strict rules had been put into place to ensure that all profits from trading found their way back to the sponsoring company. On past ventures ships' officers and even individual crew members would often supplement their income by trading directly for furs or other local items, then selling them at future ports of call and pocketing the money. On this voyage trading was to be done exclusively by the younger brother of the company's owner, and solely for the benefit of the Etches Company. No one other than the owner's representative "was allowed to trade or barter for any curiosities." This would have a significant impact on Menzies. With limited time in port, collectors often relied on trading for specimens already gathered by the locals, and clearly the mandate prohibiting this trade could severely limit Menzies' ability to perform his duties as a naturalist. He once again turned to Joseph Banks.

May I request the favor of a recommendation letter to Mr. Etches … as it may in some measure exempt me from this restriction; especially while my aims in collecting seeds, specimens and other curiosities do not interfere with the object of the voyage or in the interest of the company.[195]

Banks again interceded on Menzies' behalf and, as someone to whom few people would deny a request, was almost immediately successful. The response from the company president reveals a great deal about how Joseph Banks was viewed by the traders and merchants occupying the upper echelons of British society.

I feel duly … bound to pay (Mr. Menzies) every possible attention. I believe you are fully acquainted with the restrictions laid down in the articles … and I presume such restrictions are absolutely necessary, but in the present instance it is my full intention to dispense with them in the case of Mr. Menzies, so far as can have any tendency to be beneficial to science in general. . . and as my younger brother … is going on the voyage, I gave him orders to pay every attention to Mr. Menzies and to give him ample latitude in his pursuits.[196]

Departing from England, the two ships first traveled to the Cape Verde Islands off the west coast of Africa. The ships continued south, crossing the Atlantic to the Isla de los Estados, in the Tierra del Fuego archipelago, where they spent three weeks. Menzies had limited opportunity to collect in either location; the time in the Cape Verde was too brief, and on the ship's second stop his time was taken up by gathering the wild celery needed to combat scurvy during the next long leg of the voyage. Menzies saw the use of that plant as a temporary measure that would keep the crew healthy until their next scheduled port, in Hawaii. But that stop was not to happen. Without consulting Menzies, Colnett decided to sail directly to Quadra Island.* The captain believed he still had a healthy crew, and by starting the collection of furs as quickly as possible he felt he could maximize the profits, which would please his employer. This proved to be an extremely bad decision. Except for the wild celery taken from the Isla de los Estados, and with no opportunity to lay in supplies in Hawaii, there were no fresh fruits or vegetables onboard, so most of the crew became seriously ill from scurvy before the ships could reach the Northwest Coast of America.[197] Menzies' medical skills, with limited resources and limited assistance, were stretched thin. It was only through his extraordinary efforts that none of the men lost their lives during this phase of the voyage.

Once the ships had reached the Northwest Coast the crew acquired the necessary foods to combat scurvy and soon returned to health. This accomplished, the real work of the voyage could begin. The purpose of the expedition was the collection of sea otter skins. The ships worked from a permanent base in Nootka Sound on Quadra Island, an area already claimed

* Now Vancouver Island

by Spain. From this location, Colnett made a series of trading journeys that ranged as far north as Mount Fairweather, near the northernmost point of what is now the Alaskan panhandle. This allowed Menzies the opportunity to search the area around each harbor where they anchored as well as to barter with the natives for plants and other artifacts. Among his talents, Menzies had a unique gift for languages. On both the Northwest Coast and later, on Hawaii, he proved himself a skilled communicator with the local populations. While he was on shore, usually accompanied by one of the women from the area, his search for plants for both Kew and Edinburgh was regularly rewarded.

Although Menzies was quite well received by the local populations, Colnett and those involved in the trading for skins were not as fortunate. The journals kept by those onboard describe a series of attempts by the native populations to steal anything possible. On one occasion the ship's anchor was stolen while the indigenous traders were onboard bartering over otter skins. On more than one occasion these incidents evolved into open warfare, resulting in deaths on both sides. This was true along the Northwest Coast, but especially true in Hawaii, where Captain Cook had been murdered only a few years earlier.

For the next two years, the *Prince of Wales* and the *Princess Royal* followed the typical pattern of those trading along the Northwest Coast. Usually around November, as the weather was turning cold in North America, the ships would set sail for Hawaii. They would remain there, resupplying with fresh food and water, and repairing any damage to the ships. While in Hawaii Menzies had the opportunity to gather plants from a completely different environment than that found in North America, and he made the most of these layovers by gathering, documenting, and cataloging his finds. In early March the ships would begin their return journey, revisiting the area between Vancouver Island and Prince William Sound on the Northwest Coast, and resuming the collection of otter skins. This continued until September 1788 when the two ships, having collected over 1,200 pelts, sailed west to Canton,[*] where the skins were traded for tea and silk.

Colnett, not wanting to return directly to England, decided to proceed to nearby Macau, hoping to find another ship that would allow him to return to the Northwest Coast for more trading. Meanwhile, the two original ships, under new captains, began their voyage home. Menzies remained with the *Prince of Wales*, which provided additional collecting stops at Sumatra, Singapore, Martinique, and Madagascar while sailing west toward the Cape of Good Hope. He arrived back in England on July 14, 1789, having been away nearly three years, and during the course of the voyage he had sent more than 100 plant samples to Banks, accompanying each with an exact description of the locality where he found it, its identity, and its habitat. These specimens eventually found their way to the British Museum.

During the time Menzies was away Dr. Hope had died, but he had kept his promise to send specimens to the Royal Botanic Garden Edinburgh. So, Menzies, soon after settling in, traveled to Edinburgh and introduced himself to Hope's successor, Professor Rutherford. Menzies was becoming recognized as a legitimate botanist. In 1790 he was made a Fellow of the Linnean Society and at the end of that year he received a warrant promoting him from

---

[*] Now Guangzhou

assistant surgeon to surgeon. He had developed a taste for adventure, and he was already looking for an opportunity to set forth again.[198]

HMS *Discovery*

Luck was again with Menzies, and less than a year after his return to England the British Government decided to sponsor an around-the-world expedition, its first since Captain Cook's third voyage in 1779. This new voyage would have three distinct objectives. The first was to confirm the existence of what is now known as Antarctica. The second was to determine whether the Northwest Passage, the fabled route around North America that would link the Atlantic and Pacific oceans, existed. The third was to finalize the agreement that had been negotiated between Spain and Britain; its terms would allow the British to trade along the Northwest Coast of America, and to use Nootka Sound as an anchorage and base. Captain George Vancouver (**Figure 6.2**) would command the expedition, which would include HMS *Discovery* and its tender,* HMS *Chatham*.[199]

George Vancouver was an Englishman of Dutch descent. The family name originates from the Dutch name van Coevorden, meaning "one from Coevorden", a small town in the eastern Netherlands, where the explorer's ancestors had lived before coming to England.[200] George had been born in 1759, and as a 14-year-old midshipman he had sailed around the world on the *Resolution* with Captain Cook. He had also been present on Hawaii during the fighting that led to Cook's death.

Figure 6.2 George Vancouver (Wikimedia Commons/PD)

Vancouver was an experienced Royal Navy officer and noted explorer. Upon completion of the expedition he was about to lead, his survey of the Northwest Coast of North America would be viewed as perhaps the most extensive exploration ever undertaken and both Vancouver Island and the city of Vancouver would later be named in his honor. He was credited with being organized and a strict believer in firm discipline. He was also described as

---

* A ship used to service or support other ships, generally by transporting people and/or supplies to and from the shore

being stubborn, having an uneven temper, and being fiercely protective of his right to control all aspects of the daily operation of the ships under his command.[201]

Joseph Banks felt that when considering a voyage of the scale about to be undertaken by the *Discovery* the inclusion of an experienced naturalist was essential. At that time in his career, and with the ear of King George, Banks was also in a position to influence who the naturalist would be. He sent a letter to his close friend John Pitt, the Second Earl of Chatham, who was then the First Lord of the Admiralty. A second letter went to another friend, William Grenville, who had recently been named Home Secretary. Their return correspondence indicated that Menzies would be invited to join the expedition.[202] Banks provided Menzies with numerous and very specific instructions regarding the collecting, care, and transport of plant specimens he was to gather. He also insisted that Menzies be given special consideration on the voyage.

Unfortunately, two of Banks' requirements would create a major rift in the relationship between Menzies and Vancouver. The first was Banks' insistence that a special frame, 8 feet by 12 feet, with glass panes, be constructed on the ship's quarterdeck* to house the botanical specimens collected on the expedition. This was because Banks, during his voyage with Captain Cook, had lost most of his collection when a storm resulted in the loss of his plant boxes that had been located on a lower deck. For Banks, raising the glass houses to a high point on deck was, based on his experiences, simply to ensure the protection of the collection. But for Vancouver this was a major intrusion into a space reserved for the ship's command and would be a source of irritation to him throughout the voyage.[203]

A second demand would cause even greater problems between the captain and the naturalist. All officers participating in an exploring voyage traditionally kept a journal. At the end of the voyage, the commander of the expedition would collect the individual journals, from which he would create the official summary of the expedition. But notes written by Banks before the *Discovery* had even set sail indicate who he felt had rightful access to Menzies' journals.

> [Menzies was] … to deliver his journal to his employers on his return provided that [if it] was thought proper for publication, he should [then] be allowed to publish it for his own benefit.

Although the term "employers" is ambiguous, it is clear that Banks wanted complete access to all records or collections, and to everything written and collected by Menzies. It's clear that in Banks' mind his note was intended to mean that Menzies' records were to be handed over to the Home Office and not shared with anyone else, including Captain Vancouver.[204]

This request went from Banks to Grenville. The following day, in a letter sent directly to the Admiralty, the Home Secretary stressed the need for Vancouver to provide every assistance to Menzies, including the location of the plant boxes on the quarterdeck, the provision of fresh water for the plants, and a small boat and crew for Menzies if needed, and also—which

---

* The quarterdeck is a raised area towards the stern of a sailing ship, from where the captain commands their vessel.

was what would prove to be the most problematic—an insistence that all communication was to go from Menzies to the Home Office without oversight by the ship's command. The Admiralty conveyed these directions to Vancouver.[205]

Menzies never saw any of the directives generated by Banks, directed to the Home Secretary, passed to the Admiralty, and given as commands to Vancouver. In fact, they were marked SECRET, and were seen by no one outside the chain of command. Perhaps the issues which were to arise later in the voyage would not have done so if Vancouver had known that Menzies knew nothing of the unwelcome arrangements foisted upon him. Vancouver undoubtedly saw Menzies as "Banks' man" and a "protected person." Banks was rich, held considerable standing within the scientific community, and was a personal advisor to King George. It did not help that Banks and Grenville had persuaded the Admiralty to issue the instructions to Vancouver; he would have seen the entire process as undermining his right to control his own ship.[206] The relationship between the captain and the naturalist was not off to a good start.

Surgeon or naturalist

Originally, Menzies had requested an appointment as the ship's surgeon. He explained in his journal:

> I requested leave … to go out as surgeon of the *Discovery*, promising at the same time that my vacant hours from my professional charge, would be chiefly employed … in making such collections and observations as might elucidate the natural history of the voyage.[207]

As it seems that Vancouver objected to Menzies' appointment, a compromise was reached; within the *Discovery*'s logbooks, Menzies is listed as a supernumerary officer. This term was used at the time to indicate an auxiliary officer, or one who is added in excess of the normal or necessary number needed.

Although Menzies had received his surgeon's warrant, the two ships were already fully staffed with medical personnel. The *Discovery*'s personnel included Dr. Cranstoun, along with three surgeon's mates, Hewitt, Mill and Mears, and the *Chatham* had Dr. Walker and Dr. Nicholl. So, Menzies had no reason to believe that he would be called upon to perform any medical duties. In addition, the specifics of what he was to do were also quite vague.

Before the voyage began, Menzies, Banks, and Vancouver had all appeared anxious regarding Menzies' position within the ship's officers' hierarchy, and each of the men responded in a different way. Menzies used every opportunity to emphasize his role as naturalist; he was not simply an "excess" officer but was charged with very specific duties. Initially, Vancouver seems to have accepted this role when he wrote:

> Botany … was an object of scientific inquiry with which no one of us was much acquainted; [however] Mr. Archibald Menzies, a surgeon in the Royal Navy, who had before visited the Pacific Ocean … had, doubtless,

given sufficient proof of his abilities to qualify him for the station it was intended he should fulfill.

Banks attempted to justify the need for Menzies within the ranks of the expedition by sending numerous detailed lists of the duties he was to perform, stressing the need for Vancouver to support him in every possible way. Copies of the letters were once again forwarded to the Admiralty to ensure that Vancouver understood that these were directives, not suggestions. Ultimately, in fact, it was Vancouver who would determine the role Menzies would play while under his command. But this would take place well into the voyage and half a world away.[208]

The voyage begins

The two ships left Falmouth on April 1, 1791, and the first stop was the island of Tenerife, off the north-west coast of Africa, where Menzies was able to begin his botanizing. But this initial effort would be short-lived. Upon leaving England, Menzies occasionally performed medical duties as required. His voluntary acceptance of medical responsibilities was quite different from an official medical assignment, and he seemed to perform these periodic tasks willingly. But shortly after leaving Tenerife this changed, as Captain Vancouver fell seriously ill. He had been experiencing health issues from the start of the voyage, but this relapse was different. He requested that Menzies exclusively provide his care and treatment; this meant that Menzies' medical role was now official, and he began his treatment of the captain, which would continue throughout the voyage.

The ships continued traveling south, along Africa's Atlantic coast, and arrived at the Cape of Good Hope. They were to spend five weeks in South Africa, and Menzies was hopeful that his collecting could begin in earnest. But this time an outbreak of dysentery on the ships while still in harbor put a stop to his botanical efforts. Then, upon leaving South Africa, the two ships sailed east, toward Australia, but the cases of dysentery continued to mount in both number and severity, and one of the marines died from the disease before they could reach Australia. While still en route, Vancouver once again increased Menzies' medical duties, placing him in charge of the sick bay.

The expedition's arrival at King George's Sound, on the south coast of Western Australia, is the first recorded visit to the harbor by a European. With a long-anticipated layover, and now with the dysentery under control, Menzies was finally able to work at his original assignment; he spent many productive and effectual days exploring and collecting in both Australia and, later, New Zealand. After supplies were loaded for the next leg of the voyage, the *Discovery* and *Chatham* set sail for Tahiti. It would be at this fabled port of call that an incident occurred that would seal the captain's fate upon his return to England.

Menzies was not the only supernumerary participating in the expedition. Just before leaving Falmouth, he describes a visit to the ships by Baron Camelford and his son, Thomas Pitt. Camelford had been a member of the House of Commons until raised to the peerage in 1784, which entitled him to sit in the House of Lords; he was at the apex of both social and political society in Great Britain. William Pitt the Elder, the former prime minister, was

Camelford's uncle, and William Pitt the Younger, the future prime minister, was his cousin. Thomas was the most prominent of a group of 15 young gentlemen whose acceptance onboard had been arranged by their aristocratic and influential relatives. Although the boys' inclusion on the voyage was seen as the maritime equivalent of the Grand Tour, they were to serve on the expedition as ordinary seamen. But none of them had experience as sailors, and few aspired to a career in the Royal Navy. Most importantly, Vancouver viewed their presence as yet another thorn in his side.[209]

Under the lash

At that time, the original inhabitants of South Africa, Australia, and New Zealand were largely scattered and tended to live some distance from where the expedition's ships anchored. Tahiti was the first port of call where the members of the crew encountered a substantial aboriginal population that lived in close proximity to their anchorage. Captain Cook and Captain Bligh had both experienced problems associated with this idyllic locale. Although the island was noted for its abundance of food, and its beautiful and willing supply of female companionship, its natives had also become extremely skilled at stealing from visiting ships. Metal tools, and any pieces of metal that could be formed into useful implements, were especially prized.

To combat this problem, Vancouver, as was the practice of visiting captains before and after his arrival, had instituted strict rules. Access to shore, and the accompanying temptations were allowed, but strictly regulated. Trade with the natives, however, was not permitted except when conducted through the captain. But then, shortly after their arrival, Thomas Pitt tried to trade a broken barrel hoop for a young girl's sexual favors. Vancouver discovered this violation of the rules and sentenced the lad of 16 to two dozen lashes. Bent over a gun, Pitt was flogged in front of the whole crew. This was undoubtedly a great humiliation for the boy but would have even worse consequences for Vancouver at the end of the voyage.[210] This incident, although reported by Menzies in a letter, did not affect him directly, and his records indicate that his collection efforts on Tahiti were his most successful to date.

Nootka

Upon leaving Tahiti the *Discovery* and *Chatham* sailed for Hawaii. Because of his visits to the islands on his earlier expedition, Menzies was familiar with the location and the people, and while he was serving on the *Discovery*, he would visit a total of three times between 1792 and 1794. Throughout these visits the relationship between the members of the expedition and the islanders was generally good. Menzies, having mastered the language and with a genuine liking of the people, was especially successful, and was regarded with favor and even affection. In addition to his ongoing collection of plants, he detailed many of the customs, religious observances, agricultural practices, and habits for each island he visited. His anthropological observations would prove invaluable to later researchers.[211]

The first visit to Oahu was in March of 1792. It was brief, and the vessels were soon on their way to North America. Over the course of the next three years, the two ships would resume the pattern which Menzies had followed on his earlier voyage. The late autumn and winter months would be spent among the Hawaiian Islands, and each spring and summer would be spent exploring the Northwest Coast of North America. Rather than collecting seal pelts as had been the focus of the earlier Colnett expedition, this time Vancouver would be attempting to finalize negotiations with the Spanish regarding the fate of Nootka Sound, and to look for a possible western entrance to the Northwest Passage.

Vancouver began by surveying the coast along what is now Oregon and Washington. For the first time, they sailed past the mouth of the great Columbia River, and its size and location led many to think they might have found part of the Northwest Passage. Although Vancouver did not mention the sighting of the river in his journal it must have caught his interest, as he would return to explore this area on another journey. The *Discovery* and *Chatham* continued north to the Spanish settlement at Nootka, on the western coast of what is now Vancouver Island. This small trading village had been established to clearly mark Spain's claim to the whole Northwest Coast as an addition to its well-established settlements throughout Nueva California. It was also seen as a buffer to stop a Russian intrusion from the north. However, the newest and most immediate threat to the Spanish was now from Great Britain. Vancouver had been told that London wanted Spain to "relinquish" Nootka and the surrounding area, and he had been given instructions to accept the settlement on behalf of the Crown. But this directive was delivered with no specific instructions.

Upon arriving at Nootka, Vancouver was warmly greeted by Don Juan Francisco de la Bodega y Quadra, who commanded the Spanish settlement. But before negotiations could begin, Quadra became quite unwell, so once again, Menzies was asked to act in his role as ship's surgeon and treat him. Fortunately, Menzies' efforts were effective, and Quadra was soon recovered enough to meet with Vancouver. Negotiations began in earnest.

The discussions remained friendly, yet little headway was attained. Quadra was willing to turn over a small portion of the island and Nootka Sound for the exclusive use of the British. But Vancouver wanted more; his plan was to take possession of the whole region, exclude the Spanish from the area and make it a permanent British settlement. As it seems that Quadra's directions from Madrid were as vague as Vancouver's from London, in the end—after all of the personnel, effort, and drama that were part of this visit—the two suspended their negotiations without reaching a settlement and parted as friends.[212]

All indications are that Menzies was a valued member of the expedition. He had been selected by Vancouver as his personal physician, and in addition the decision to have him treat the head of the Spanish garrison point to his prominence and position. Records show that Menzies was being asked to accompany Vancouver on all important meetings, such as the negotiation sessions at Nootka. Although still listed as a supernumerary officer, Menzies had clearly been recognized by Vancouver as one of the most valued members of the crew. He was dependable, skilled, and intelligent. But this recognition was soon to come with a price.

By September of that year, Dr. Cranstoun became ill and had to be sent home. Two days later, Vancouver appointed Menzies to take his place. He would now be responsible for the supervision of the medical staff on both ships, and assume major medical responsibilities himself, in addition to his primary role as naturalist. Vancouver asked Menzies to "voluntarily" accept this new assignment, but should he choose not to do so, he would need to decline the assignment in writing. Menzies unwillingly accepted. Perhaps this additional assignment speaks to the captain's faith in his surgeon's medical abilities. But Menzies saw the appointment as both adding significantly to his duties and putting him firmly under the command of Captain Vancouver.[213] And perhaps he was right.

Northwest Passage

Before returning to Hawaii, Vancouver decided to continue his survey of the Washington and Oregon coasts. This time he chose to explore the Columbia River, sending the *Chatham* ahead as it was the smaller of the two vessels, but bad weather soon forced both ships out of the river. Concerned that the weather was not going to improve, he commanded the ships to turn away and sail west toward their winter berth in Hawaii. It was on this return trip that Menzies would participate in an event that would gain him as much, or possibly more, recognition than all his botanical discoveries.

It was January 1794, and Menzies decided he wanted to climb Mauna Kea, the highest mountain in Hawaii. **(Figure 6.3)** In addition to meeting the challenge provided by the climb, he wanted to expand his collecting activities into elevations he had not before explored. He initially climbed two slightly smaller peaks as preparation for the ultimate ascent, and finally, in February 1794, he felt ready to make the climb. Menzies and two shipboard companions crossed lava fields that shredded the soles of their shoes, fought their way through dense woods, circumvented

Figure 6.3 Mauna Kea (Wikimedia Commons/PD)

active volcanoes, and braved extremely cold temperatures, while continuing to collect and catalogue plants along the way. On February 14 the three men finally reached the summit, where they spent two days. This was the first known ascent of the mountain by a European, and this feat would not be repeated until another naturalist, David Douglas, made the climb in 1834.[214] As fate would have it, the successful climb was broadly reported. Menzies became a hero in the public eye. For all of his skilled scientific efforts, his ascent of Mauna Kea brought him more fame and admiration than any of his previous professional work.

Following the winter of 1793–94, the expedition returned again to the Northwest Coast of North America for their third season of surveying. Vancouver began at the northernmost point of Cook Inlet, near the present-day city of Anchorage. He then began systematically working his way southward in an attempt to locate an outlet for the Northwest Passage. By this time most of the surveying had been delegated to junior officers, as Vancouver's health had deteriorated to the point that he was forced to remain in his cabin for days at a time. He was experiencing difficulty breathing, and his face and body were becoming increasingly bloated.

During the previous season the expedition had visited several Spanish ports along the coast of Nueva California. They had sailed to Santa Cruz and San Francisco, and as far south as San Diego. Now, following their extensive surveying and search for the Northwest Passage along the Northwest Coast of the continent, the two ships, sailing south and taking advantage of the mild climate, stopped at Monterey to rest and lay in supplies for what lay ahead; the difficult journey toward Cape Horn would be seriously challenging. Vancouver and his men had spent three years creating what were to date the most advanced charts and maps of the entire Northwest Coast. Of their efforts it was said that they had completed one of the greatest and most difficult voyages of discovery that had ever or would ever be undertaken.[215] The captain and his men had faced death, disease, hostile locals, inclement weather, and an insubordinate crew, yet had still achieved most of their objectives. But to Vancouver's disappointment they had also failed to find any indication that the Northwest Passage existed anywhere between Alaska and the southern tip of Nueva California.

Monkey puzzle tree

After taking on supplies, repairing the ships, and giving the crew a chance to rest at Monterey, the *Discovery* and *Chatham* began their long voyage home. With unfavorable winds and oppressive heat, the journey became more and more difficult. In January 1795 they finally reached Cocos Island, off the western coast of Costa Rica, where they were able to secure food and water. From Cocos, the ships sailed south-southwest to the Galápagos, where Menzies again was able to focus on plant collecting. However, the islands failed to provide a wealth of plants, Menzies describing them as the "most dreary, barren and desolate country ever beheld." [216]

When the ships entered harbor at Valparaíso, Chile, they were greeted by an invitation from the governor, Don Ambrosio O'Higgins de Valenar,* to visit him in Santiago. Vancouver selected six of the ships' officers, including Menzies, for the 90-mile ride to the nation's capital. They received a gracious welcome from Don Ambrosio, who had arranged a grand banquet. At the close of the meal Menzies, ever observant and curious, collected some unfamiliar nuts left on the table. Carrying them back to the ship, he planted them in the glass cases, where they germinated. By the end of the voyage he had five strong seedlings, which he gave to Joseph Banks at Kew.

Menzies has been described as "one of greatest Scottish botanists, explorers and travelers." Through his extensive travels and avid collecting he introduced hundreds of new plants to England; 25 plants found along the Pacific coast of North America have been given his name, and 19 indigenous to Hawaii. But it was the chance discovery at the end of the Chilean governor's banquet for which he is best known. This was the *Araucaria imbricata*, and one of the original seedlings, planted at Kew in 1795, lived until 1892.[217]

This variety of evergreen ultimately grows to a height well above 100 feet. Its bark has been

---

* Born Ambrose O'Higgins in Ireland; his aristocratic family having been evicted by Oliver Cromwell's troops, he became an adventurer.

described as looking like a reptile's skin. Its branches do not emerge until near the top of the tree, forming an oddly shaped silhouette. This odd and quirky tree quickly became the most sought-after plant for British gardens. In about 1850 Sir William Molesworth, the owner of a garden in Cornwall, was showing his mature specimen to a group of friends. One of them, remarking on its height and shape, remarked, "It would puzzle a monkey to climb that." Until that point, no other popular name for the tree had existed. From then on, it became known as the monkey puzzle.[218] By the time Menzies' last surviving seedling had died, the monkey puzzle was a common sight in perhaps thousands of British gardens. **(Figure 6.4)**

Arrest and court martial

Leaving Valparaíso, the ships rounded Cape Horn. This appears to be as close as the explorers would get to Antarctica,

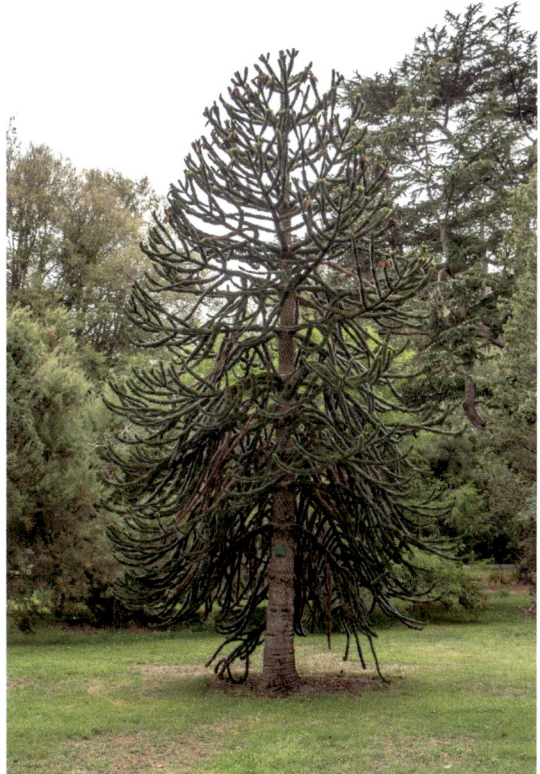

Figure 6.4 Monkey Puzzle Tree (Wikimedia Commons/PD)

seeming to ignore what had been portrayed as the most important of the three directives they had been given prior to departure. The decision to eliminate the more southerly route might, however, have been due to the condition of the ships. The storms they encountered as they rounded the cape caused such damage that they reportedly limped into the harbor on St. Helena, where both ships required extensive repairs.

Following a period of heavy rain after leaving St. Helena, Menzies discovered that the glass houses had been left open and much of his collection was ruined. It turned out that the sailor who had been delegated to look after the frame had been moved to another assignment without Menzies being told. Menzies felt that his carelessness with the frame was a dereliction of duty and demanded that the man be punished. Vancouver refused, as the sailor had been following an order given by a superior officer. The captain also reminded Menzies that he was still waiting for him to produce his journal, as the other officers had done, and that he intended to impound all journals and logs prepared by "the officers, petty officers and gentlemen onboard the *Discovery*." [219]

Perhaps Vancouver was attempting to assert his authority in spite of the clear directives presented to him by the Admiralty. He could also have been following past practice by impounding all journals and logs until the official summary of the expedition could be written.

Regardless, Menzies felt his directive from both Banks and the Home Secretary superseded the captain's order, and he refused. As the quarrel had taken place publicly, Vancouver undoubtedly felt he must respond in kind. Menzies was placed under arrest, confined to his cabin for the remainder of the voyage, and told he would be court-martialed upon their return to England.

On October 2, 1795, the *Discovery* was safely docked on the Thames near Deptford. Now back in London, Menzies must have experienced a mixture of emotions regarding the voyage. Forced to assume duties as chief medical officer in addition to his responsibilities as naturalist, it must have been a conflicting and difficult time for him. Yet he had served honorably in both capacities. In the course of the five-year voyage only four individuals had died; a marine had died of dysentery early in the voyage, before Menzies had officially assumed his combined position; one seaman had been poisoned by eating tainted mussels, and two had drowned. This low death toll was a remarkable accomplishment by Menzies, as on voyages of this type the loss of life was usually much higher.

As a naturalist, Menzies had studied the flora of South Africa, Australia, New Zealand, Hawaii, the smaller Pacific islands from Tahiti to the Galápagos, the entire Northwest Coast from Alaska to the southern tip of Nueva California, and the Pacific coast of Chile. Although many of his seedlings had been lost because of the incident with the glasshouse, he had dozens of boxes of plants and packets of seeds that would greatly enrich the national collection.

But ironically he was under arrest and facing court-martial. He was not allowed to leave the ship, nor unload his collection, but he could send a letter. He immediately contacted Joseph Banks.[220]

Vancouver's return

Menzies was asked by Banks to submit an apology to Vancouver, and he complied. At the same time, Vancouver was pressed by the Secretary to the Board of Admiralty, through Banks' intervention, to drop the proceedings for the court-martial. Vancouver agreed and withdrew his request. At last Menzies was free and allowed to go ashore. The feud between Menzies and Vancouver was permanently put to rest, and neither spoke ill of the other ever again. But one of Vancouver's earlier decisions would have a more severe effect on him.

Thomas Pitt's flogging for his indiscretion in Tahiti did not seem to have made much of an impact on his future behavior. At some point on the voyage, he was found wrestling on the quarterdeck with another of the young men in his group. As the men wrestled, Pitt broke the glass on the binnacle, a waist-high stand that holds delicate navigational instruments vital to steering of the ship. He received another flogging. Finally, he was found sleeping on watch. For these irresponsible and dangerous actions, Pitt was placed in irons. When another ship could be located he was transferred to it, and sent to England in disgrace.

All of this was appropriate discipline for the era. However, no one on the expedition could have known that Thomas's father had died in 1793. As the first-born son, Thomas Pitt was now Baron Camelford and a Member of the House of Lords. By the time Vancouver returned to England, Baron Camelford had been telling his version of events to his allies for the two years since his return. Among his circle was his cousin, William Pitt the Younger, now Prime Minister. A series of increasingly vitriolic letters from Thomas Pitt to Vancouver began the attack. This was followed by a series of newspaper articles painting Vancouver in a most unfavorable light. The Pitt campaign against Vancouver finally culminated in Thomas's brother, Charles, encountering Vancouver on a London street. There Charles beat him with his cane until onlookers, afraid that he would kill Vancouver, had to pull him aside. Newspaper articles reported the case, but the Camelford faction had the press firmly on their side. A potential court case filed by Vancouver was repeatedly delayed. Meanwhile Vancouver's health continued to deteriorate, and within a year he was dead. No charges were ever filed.[221]

The final days

Menzies, now 41 years old, had completed two extremely successful round-the-world collecting voyages. He was a valued member of the scientific community and had Joseph Banks as his friend and advisor. His vast collection of plants would forever alter the British landscape. His hard work had placed him at the top of his profession. It would appear that additional success must lie in his future—but in fact it would be these two voyages, made during the first half of his life, on which his fame would rest. Although he would continue to be respected and valued by the scientific community, his return from the *Discovery* would mark the beginning of his fall from public life, and upon his death he would soon be forgotten.

There were two major factors which caused this turnabout in Menzies' personal and professional position. First was the change in his relationship with Joseph Banks. In spite of Banks' past support and their ongoing mutual interests, a coolness seems to have crept into their relationship. It could that this was a result of the charges leading to Menzies' court-martial. Perhaps Banks was tired of having to intercede repeatedly so directly on his protégé's behalf. Whatever the reason, after Menzies' return the relationship was never the same. Both of the round-the-world voyages had been through Banks' recommendation. Another such endorsement never came.

Second was Menzies' reluctance to publish any of his findings. Although he had kept copious records and faithfully recorded each aspect of the voyages in his journal, he contributed no papers or articles to the literature. Much of his work had to be left to others, notably William Hooker, to publish, but these publications were not completed until well into the 19th century. There is no doubt that the lack of any major published work by Menzies seriously affected his position within the scientific community and his lack of visibility among the general population.[222]

It is also probable that the two reasons were firmly linked. In March 1796 Vancouver was ordered to prepare his journal for publication. He learned that there was a plan to include Menzies' journals in the project, just as Banks' journals had been included in Captain Cook's publication. Based on their earlier conflict Vancouver protested, and he was successful; Menzies' writing would not be included.

Banks then contacted Menzies and directed him to begin work immediately in order to have his own journal published ahead of Vancouver's. But Menzies needed employment, and he needed an income, so he resumed his service with the Royal Navy, at the same time attempting to ready his journals for publication. He made progress, but not quickly enough, and he wrote to Banks:

> I received your kind letter this morning and return your kindly admonitions and solicitations respecting the finishing of my journal before Captain Vancouver's is published. It is what I most ardently wish, for more reasons than one, and therefore have applied to it very close. . . (I) assure you that I will continue with unremitting application until the whole is accomplished.

However, it was not to happen. When Menzies failed to publish ahead of Vancouver, the relationship between Banks and Menzies cooled even further, Banks even refusing to support the publication of Menzies' writings when they were ultimately completed.[223]

When Menzies resumed his military service, it was a quiet and fairly unremarkable time. His naturalist duties and collecting had ceased, and his responsibilities were solely medical. He once again served as a surgeon, sailing for most of the next three years on the *Princess Augusta*. He now seemed more focused on his medical credentials, and in 1799 he was awarded a medical degree by the University of Aberdeen. He then returned to shipboard duties, serving on the *Sans Pareil*, spending much of his time back in the familiar haunts of the West Indies. He continued in the Royal Navy until 1802. He then mustered out at half-pay, married, and began a private medical practice in London, which lasted for another 20 years.[224]

In 1826 he retired and moved to the Notting Hill area of London. He spent his time sorting his collections and experimenting with plants in his back garden. He was reported to be a welcoming host who enjoyed telling stories about his various travels. **(Figure 6.5)** It seems that his favorite tales were about Hawaii, where he had often been invited ashore to treat injured natives. It was Hawaii where he was long remembered as "the red-faced man who cut off the limbs of men and gathered grass." [225] He also enjoyed mentoring the next generation of naturalists, and corresponded regularly with a number of them, including David Douglas and John Scouler.

To sum up, Archibald Menzies, one of the early medical explorers, had come from a family of tenant farmers. His father had been a crofter with a house, a small plot of land, and some shared pastureland that he worked collectively with his neighbors and extended family. The male members of the family all held second jobs as nurserymen to supplement the meager returns from the croft. Yet the educational system in Scotland, greatly advanced during the Enlightenment, had ensured that Menzies was well educated, and the academic promise that he exhibited had allowed him to secure a place within the University of Edinburgh's medical program.

Figure 6.5 Archibald Menzies (Wikimedia Commons/PD)

Archibald Menzies (1754–1842)

The Acts of Union had merged the Scottish and English military units, and Menzies, upon leaving his studies at Edinburgh, had been provided by the British military with a position as assistant surgeon with the Royal Navy. The combination of his surgical experience with Edinburgh's emphasis on humanities and science in addition to its traditional medical studies, and Great Britain's overwhelming interest in natural history, allowed him to be selected for two round-the-world voyages. His background, education, and access to opportunities not available in earlier times had provided him with the knowledge and expertise to discover, collect, catalogue, and bring back hundreds of hitherto unknown plants, which resulted in a complete transformation of the British gardens and countryside. His work also provided significant amounts of material for the advancement of botanical science, and enhanced collections across the United Kingdom.

Archibald Menzies' wife, Janet, died in 1836. He lived alone for the remainder of his life, dying in London on February 15, 1842, one month short of his 88th birthday. John Joyce Keevil writes:

> He died at Ladbroke Terrace, leaving his beloved specimens to the various national collections. He himself lies in a depressing London cemetery, but the whole countryside of England and Scotland keeps his memory green.[226]

7

Sir John Kirk (1832–1922)

When you encounter slavery you may choose to look the other way
but you can never say again that you did not know.

–William Wilberforce

By the 18th century slavery and the triangular trade had become a major economic mainstay for Britain, especially in the cities of Bristol, Liverpool, and Glasgow. The ships involved in slavery would set out from Great Britain loaded with trade goods bound for West Africa. On the ships' arrival along the African coast some of these items were sold and some were exchanged for enslaved people captured by coastal groups who had traveled into the interior of the region to collect other Africans. The British, having exchanged their manufactured goods for human cargo, then sailed to the West Indies or to the American colonies, where the captives were sold to be used on the British-owned plantations. Part of the profits went towards the ships being loaded with the products of the slave labor, such as cotton, sugar, and rum, and the traders completed the third leg of the triangle by returning to Britain, where these goods were sold for even more revenue. The pattern was repeated hundreds of times each year, with British merchants earning huge profits at each stop along their ships' journeys.

But by the end of the 18th century, despite the massive positive economic impact on individuals and on the country as a whole, the anti-slavery Members of Parliament were becoming a major political force. Their leader, William Wilberforce, had become the primary spokesman for the abolition of slavery in Britain. In 1807 his Slave Trade Act abolished the trading of enslaved people within the British Empire. In 1808 Britain established the West Africa Squadron, utilizing the Royal Navy to suppress the slave trade by patrolling the West African coast. Ships of the squadron would stop vessels that might be transporting enslaved people, regardless of the flag they were flying. If their suspicions were founded, the vessel was claimed as a prize for the British government, and the captives were transported to Freetown, the capital of Sierra Leone, where they were released. Between 1808 and 1860 the squadron captured 1,600 slave ships and "freed" 150,000 Africans. But the cost was high. During that same period over 1,500 British sailors died, most of them from disease, while serving with the West Africa Squadron.[227]

Although the numbers of captured ships and freed Africans seem large, the efforts of the West Africa Squadron in actuality produced only a marginal impact. The number of ships captured, and number of enslaved people freed, represented only about 6 percent of the transatlantic slave trade. Also, a large number of captives were being collected by slavers who had traveled from the Arabian Peninsula. The enslaved people they had captured were transported via caravan or river routes to Portuguese settlements in East Africa, so the West Africa Squadron had no impact on that trade.[228] Organizations such as the African Association had long theorized that slavery could not be eradicated through efforts such as those attempted by the West Africa Squadron, so as mentioned in Chapter Five, they offered a completely different, two-pronged, approach for the eradication of slavery. First, they proposed replacing the trading of enslaved people with goods that were morally acceptable while generating equal or greater profits for both the Africans and the British merchants dealing in human traffic. Second, they would Christianize the pagans and the followers of Islam. It was believed that once these individuals had accepted the Christian faith they would understand that the capture and selling of their fellow humans was morally unacceptable. It was felt that only by introducing these measures could the practice of slavery be ended. This approach would be the focus of what would become John Kirk's introduction to Africa.

The minister's son

On December 19, 1832, John Kirk was born in the village of Barry, about 10 miles east of Dundee. His mother and father had four children. Of these, Alexander, three years older than John, became a successful engineer credited with numerous inventions. He became director of Scotland's largest ship-building firm and eventually became quite wealthy. More importantly for his brother John's future, early in his career Alexander had developed a close friendship with the missionary explorer David Livingstone.[229] John was the second born, and was followed by Elizabeth and James. Elizabeth died in her twenties of an unknown illness. James was quite sickly throughout his life, emigrated to South Africa, and also died at an early age.

Initially, the Kirk family was quite wealthy. John's father was a minister serving Barry for the Church of Scotland. When John was only a few years old, his father received a significant promotion. This involved not only moving to a much larger congregation and a major increase in salary, but also into a fine house provided by the Church of Scotland. But this was a time of great unrest within the Church which, as mentioned in Chapter Two, had become divided between the Moderates and the Evangelicals. John's father supported the Evangelicals, and in 1843 he joined approximately 400 other like-minded ministers in forming the Free Church of Scotland. This decision caused the Kirk family to lose both the substantial salary and their spacious family home, and they were forced to move into a small rental property in the coastal village of Arbroath. Losing virtually everything was undoubtedly a severe blow to the family, but there was a positive outcome. John's father was no ordinary preacher. He was also a scholar and amateur botanist. With time on his hands and with little money in his pocket, he dismissed the children's tutor and began teaching them himself. Under his father's tutelage

John became a keen student of botany and an expert in identifying the various plants found along the Scottish coast.

Well prepared through his father's tutoring, John was admitted to the University of Edinburgh. For the first two years he studied the classics,[*] history, and geography. In his third year he was accepted into the School of Medicine, where he proved to be a brilliant student. While at university he continued his study of botany, studying under John Hutton Balfour, who was at the height of his career, serving as Chair of the Department of Botany, Keeper of the Royal Botanic Garden Edinburgh, and Her Majesty's Botanist, while also serving as Dean of the Faculty of Medicine.[230] In 1852 Kirk's background in geography and botany, no doubt aided by his association with Balfour, was sufficient to gain him a Fellowship of the Royal Botanical Society. This was an unusual honor for a student, especially one still two years away from graduation.

After receiving his medical degree in 1854, Kirk became one of seven applicants accepted for a year's residency at Edinburgh's prestigious Royal Infirmary. This placed him within a group which would be recognized as one of the School of Medicine's most highly esteemed graduating classes. **(Figure 7.1)** Among the seven selected, John Beddoe would become recognized as one of the most prominent ethnologists in Victorian England. Patrick Watson would emerge as an eminent 19th-century Scottish surgeon and a pioneer of anesthetic development. David Christison, after a successful career as a surgeon, became one of Scotland's most noted archeologists. Joseph Lister became the pioneer of antiseptic surgery who solved many of the problems of surgical infection.[231]

Figure 7.1 Fellow residents
(Wikimedia Commons/PD)

---

[*] Ancient Greek, Latin, ancient history and ancient art + architecture

The Crimean War

The beginning of the Crimean War in 1853 corresponded with Kirk's residency, and in 1855, after completing his year at the Royal Infirmary, he volunteered for service and sailed for Turkey. He was joined by John Beddoe and David Christison. When the three newly qualified doctors arrived at Scutari, they were deeply disappointed to discover that most of the patients they were to treat were suffering from cholera. They had expected their work to be treating wounds sustained through military action. Although treating this disease was not what Kirk had planned, the experience would prove valuable to him later in his career.

Initially the doctors found little to do beyond their minimal hospital responsibilities, and the three traveled extensively including visits to some of the Greek islands, allowing Kirk to pursue his interest in botany.[232] The plants he gathered were shipped to Professor Balfour in Edinburgh, and also to a new contact he had developed, William Hooker, the Director of London's Royal Botanic Gardens Kew. After the three doctors had lodged numerous complaints related to their inactivity, they were ultimately transferred to Erenkevi Hospital in the Dardanelles. Assigned a full caseload of those injured in combat, Kirk still found time to learn Turkish and to study the Muslim religion and customs.

But in the end the Edinburgh doctors were to spend very little time treating the wounded, as in 1856 the Treaty of Paris ended the war earlier than expected. Following his time at Erenkevi, Kirk returned to London. But he had enjoyed his experience in Turkey and Greece and wanted to continue his travels. So, obtaining his discharge from the medical corps, he signed on as the physician for a tourist group visiting Italy, Egypt, Palestine, and Syria. While traveling he continued collecting plants on behalf of both Edinburgh and Kew and took the time to learn Arabic. Upon his return to England in 1857 he wrote several articles related to his travels and collecting and had them published in *The Transactions of the Edinburgh Botanical Society*.[233]

John Kirk was now 26 years old. (**Figure 7.2**) Although he had traveled extensively and had been to war, he was uncertain of the direction he would like his life to take, so he again turned to his mentor from his university days. Balfour reminded him that he was as accomplished in natural science as he was in medicine, and suggested he pursue an academic career. A professorship in

Figure 7.2 Sir John Kirk
(Royal Geographical Society)

botany was open at Queens University in Ontario, and Balfour assured Kirk that with his, Balfour's, endorsement it was his for the asking. But this was not to be. Kirk would not go to Canada, and he would not become a university professor. A series of events were about to take Kirk's life in a completely different direction, and the decision he was about to make would forever link his name with perhaps the most well-known explorer in African history.

David Livingstone

David Livingstone was born into a poor family in Blantyre, Scotland, and by age 14 he was working as a mill hand to help support his family. Wanting more out of life, he completed the entrance examination for the University of Glasgow Medical School and was accepted. He studied for two years, but then left Glasgow for London without completing his degree. He applied to join the London Missionary Society (LMS) and was accepted for missionary training. He also enrolled in the Charing Cross Hospital Medical School to complete his medical training. Finishing his training in both medicine and theology, he began attending presentations by Robert Moffat, a Scottish missionary recently returned from working with the Botswana people, and Livingstone, based on his interactions with Moffat, focused his missionary goals toward working in Africa. The LMS agreed, and he was soon on his way to South Africa. Shortly after his arrival he married Mary Moffat, Robert's daughter. Over the course of her short life she would accompany him on his numerous missionary postings and exploring journeys.

For his first ten years in Africa Livingstone served in missions in South Africa and Botswana. In 1851 he came to the conclusion that: "the best long-term chance for successful evangelizing was to explore Africa in advance of European commercial interest by mapping and navigating its rivers … deep within the continent's interior." [234] This decision would move him from the realm of obscure missionary into the sphere of world-renowned explorer. Livingstone traveled to the area where the current countries of Zambia and Angola share a border, and began following the Zambezi River upstream. He journeyed west, traveling overland to ultimately reach the Portuguese city of Luanda on the Atlantic Ocean. He then retraced his route to

Figure 7.3 David Livingstone (Royal Geographical Society)

the Zambezi and followed it until he reached the Indian Ocean. On this leg of his travels, he became the first European to see Mosi-oa-Tunya (the smoke that thunders), which he renamed Victoria Falls. On this journey, traveling from coast to coast, Livingstone became the first European to transverse south-central Africa from the Atlantic Ocean to the Indian.

The great promoter

By the end of this remarkable journey, Livingstone had come to believe that his calling was for exploration rather than for proselytization. He strongly felt that the establishment of trading centers, along with religious missions, near Africa's many navigable rivers should be the LMS's primary focus. Like the members of the African Association, he believed that conversion to Christianity, and the introduction of legitimate and moral options for trade, would create a society in the image of that found in Great Britain. His motto, now inscribed on the base of his statute at Victoria Falls, became "Christianity, Commerce, and Civilization." [235]

By the time when John Kirk was introduced to Livingstone, the renowned missionary-explorer was returning to England after being in Africa for 15 years. **(Figure 7.3)** By 1857 Livingstone's accomplishments were legendary. He was awarded the Royal Geographical Society's Gold Medal for his "discoveries, enterprise and bravery," and he was elected to the Royal Society.[236] No one could argue that these tributes and awards were not deserved. However, in addition to being courageous and resilient Livingstone was also remarkable in his abilities of self-promotion. Upon his return to Great Britain he immediately published a book, *Missionary Travels and Researches in South Africa*, relating his trans-African travels and his various discoveries. The royalties produced by the sales of his book supported him and his growing family, allowing him to concentrate on raising money for his return to the Zambezi, which he vigorously pursued.[237]

The Royal Geographical Society's president, Sir Roderick Murchison, **(Figure 7.4)** approached Livingstone about giving a series of talks regarding his ongoing plans for Africa. These presentations drew huge crowds throughout England, Scotland, and Ireland. Those who heard his message, "Christianity, Commerce, and Civilization," were struck by his passion and conviction. Among those who heard him speak on multiple occasions was John Kirk. Livingstone impressed him, and

Figure 7.4 Sir Roderick Murchison (Royal Geographical Society)

others in attendance, with his sincerity and his spirit of self-sacrifice. His listeners heard that he had answered God's calling by devoting his life to converting Africans to Christianity and to abolishing slavery throughout the continent.

Livingstone's consistently delivered message paid great dividends. The British Parliament awarded him £5,000 to explore the Zambezi and its tributaries. His former medical school in Glasgow provided an additional £3,000. Money given generously by those attending his lectures further swelled the coffers. Those who heard him speak also volunteered in large numbers to be part of this world-famous explorer's next expedition. But it was John Kirk that Livingstone selected to serve as botanist on what would become known as the Second Zambezi Expedition.

Kirk had only seen Livingstone in England, and everyone who met the vaunted explorer away from his travels described the same image; here was a man who was deferential, modest, committed, and obviously brave. Various authors, describing a complicated relationship between David Livingstone and John Kirk, note that they were remarkably alike. Moorehead writes:

> There was a complex relationship between these two. Although Kirk was much younger than Livingstone their backgrounds were very similar. Kirk too had come from a religious household in Scotland. He had also taken his degree in medicine and had traveled abroad to satisfy a craving for adventure.[238]

But other authors portray two quite different personalities. Liebowitz, in describing the relationship between the two men, wrote: "(Both) showed great tenacity as an aspect of character that allowed [them] to succeed where others had failed." But where Kirk was calm, open to direction, and introspective, Livingstone is described as having "little patience for his compatriots … made many enemies … was moody … [and] would become so completely immersed in his own thoughts that he cut himself off almost completely from his companions." Dr. Oliver Ransford, one of Livingstone's biographers, speculates that he was manic-depressive. There were periods of time when Livingstone was "bitter, short-tempered, morose, secretive and even paranoid." Yet at other times he could be "quite foolhardy and unrealistically optimistic, causing him to lead others into life-threatening situations on more than one occasion." [239] With the best-funded expedition in African history, Livingstone prepared for his return to the Zambezi River. He would be accompanied by the 26-year-old Dr. John Kirk. It was a momentous decision for the young man, and for the next 16 years his life would be inextricably linked to that of Livingstone.

Failure on the Zambezi

Livingstone was extremely well thought of throughout Britain. However, acceptance of his proposed expedition was not universal. Upon his return to England the LMS congratulated him but told him they were reluctant to support his newest plan, which they felt was "connected only remotely with the spread of the Gospel." Instead, they encouraged him to return to Africa and to serve in a more traditional mission placement.[240] But Livingstone, having decided that his calling was in exploration, had repeatedly told audiences throughout Britain what his new strategy was to be, so he felt he could not change a plan he had so publicly proclaimed. He would return to the Zambezi and put his newly formed approach into motion.

After offering him significant financial support, the Foreign Office (FO) also appointed him as Her Majesty's Consul with a "roving commission." In February 1858 the FO further defined his area of jurisdiction to be "the (entire) Eastern Coast of Africa and the independent districts in the interior." [241] This bestowed recognition and authority on Livingstone that far exceeded any other consul's influence within the British Empire. With the government's funding and the title of consul, along with the support of his former medical school and the general public, Livingstone decided he would go forward. In March 1858 he resigned from the LMS and returned to Africa.

In addition to Kirk and Livingstone, the Second Zambezi Expedition participants included: Richard Thornton, a graduate of the Royal School of Mines, hired to look for valuable minerals; Charles Livingstone, David's younger brother, in charge of evangelizing; Thomas Baines, who would supervise the coordination of supplies; George Rae, a native of Livingstone's hometown in Scotland, who would be the boat's engineer. In addition, Captain Norman Bedingfeld of the Royal Navy was chosen to command the expedition's vessel, *Ma Robert*. The ship was a side paddlewheel steam launch designed to be shipped in sections and assembled at the mouth of the Zambezi before starting upstream. According to the custom of the South African people, the boat was named after Livingstone's wife, Mary—Ma—and their firstborn son, Robert.

The group traveled on the *Pearl* from England to East Africa, but even before they rounded Cape Agulhas to enter the Indian Ocean there were signs of trouble. The first major concern was that although this was a government-sponsored expedition under the direct supervision of the Foreign Office, it was David Livingstone, a private citizen, who was in command. Livingstone would soon prove himself to be completely inept at managing a large-scale project. Second, although the terms of employment under which each member had been recruited were for a period of two years, it was estimated this was far less time than the exploration of the region would take. So, there was the potential that the entire crew could abandon the operation while still in Africa and well before the mission had been accomplished.

Third, although the Portuguese had been friendly and supportive of Livingstone during his previous time on the Zambezi, they now feared that this latest expedition was being conducted for ulterior motives. Livingstone was well known to oppose slavery, and he was on record as acknowledging that the central focus of this mission was to eradicate slavery in East Africa. Portugal had signed an anti-slavery treaty in 1842, but the Portuguese administrators did not have the appetite to enforce it. Various Portuguese governors of Quelimane, the main slaving port of what is now Mozambique, had been recalled to Lisbon for either ignoring the anti-slavery treaty or, worse, dealing directly in the trade themselves. The Zambezi was the primary slave route for the Arabs as they traveled to and from the interior, and Portugal controlled hundreds of miles of the river. In addition, enslaved people were routinely transported from its mouth to Quelimane or to the slave markets of Zanzibar, and as neither the Portuguese nor the Arabs wanted the adverse publicity that the expedition's reports to England would generate, their strategy was simply to do everything possible to impede the group's efforts.[242]

The Portuguese were also concerned that an expedition touted as an effort to find routes into the interior in order to spread Christianity was actually masking an attempt by the British to claim territory. This was a time when European nations in Africa were practicing what has been termed Informal Imperialism. Acknowledgement of a country's claim to sections of the continent was based on exercising military influence, building fortifications, and/or establishing economic dominance through creation of treaties and trade centers. So, upon receiving word of the Second Zambezi Expedition, the Portuguese immediately rebuilt outposts that had been abandoned and supplied additional troops to other garrisons. A new customs house at the mouth of the Luabo River (one of the four entrances to the Zambezi) was established to collect tariffs from the expedition. Finally, all supplies shipped to the explorers would require a Portuguese-issued permit and would have to be carried by ships flying the Portuguese flag.[243] The Portuguese were sending a strong signal that the Zambezi, and all the areas through which it flowed, would remain their territory despite Britain's attempts to the contrary.

A little over a month after leaving the Cape of Good Hope, the *Pearl* was anchored outside Kongone Harbor, the third entrance to the Zambezi River. Having bypassed the customs house on the Luabo and seeing no sign of the Portuguese at this location, Livingstone felt that this would be an ideal site for the British ships to regularly supply the expedition without being observed. So, the *Ma Robert* was assembled there, and the two vessels traveled upriver for about 40 miles. At that point the river became too shallow for the *Pearl*, and it returned to the Indian Ocean, leaving the *Ma Robert* to continue on its own.

Progress upriver, against the current, was constantly difficult. Livingstone accused Bedingfeld of testing the boilers only with coal while in Scotland, but now, with only wood to burn, they were not able to generate enough steam to effectively propel the vessel against the fast-flowing water. Other concerns arose, and the relationship between Bedingfeld and Livingstone rapidly crumbled. The matter soon came to a head, and after only a short period of time on the river Captain Bedingfeld resigned. Livingstone accepted his resignation and the ex-captain traveled at once to the Indian Ocean, where he secured passage on a merchant ship and returned to England. Livingstone assumed command of the *Ma Robert*, and Kirk, having avoided the conflict as much as possible, was now named second in command.[244]

The ship continued up the Zambezi until it reached Tete, located about 250 miles from the Indian Ocean. Progress had been slower than any of them would have liked, but apart from the incidents with Bedingfeld, the travel had generally been uneventful. The slow progress had actually worked to Kirk's advantage, as throughout this period he had time to record observations and collect and catalogue numerous plants. While at Tete, the expedition members rested, made minor repairs to the ship, and collected additional supplies. Kirk continued his observations and collections while staying in the home of the Portuguese commander and was also able to make several brief surveys of the area surrounding the city.

The party remained in Tete for a considerable period of time while Livingstone and different members of the expedition explored the surrounding territory and portions of the river that lay ahead. When the crew members returned from their various excursions, the travel upriver continued. But after journeying for only 75 miles, Livingstone suddenly sighted the Kebrabasa Rapids. Having bypassed this area of the river on his earlier travels, he was now faced with what he had to admit was an entirely unexpected obstacle. In his journal, Kirk describes giant boulders and massive waterfalls between high cliffs. The fast-moving current, with its frequent cascades and enormous rocks, continued for as far as Kirk could see. It was obvious to him that there was no way to continue by boat.

Had Livingstone been less fanatical in his belief that he was personally chosen by God to end slavery, or had he not so fervently believed, as Sir Roderick Murchison had proclaimed, that "The Zambezi River would become God's highway for bringing Christianity to the heathen," [245] he might have cut the expedition short at this point. But he was still determined to push through, and his new plan was to wait until the rainy season. With the higher water he assumed they could bypass many of the obstacles and reach the smooth water which must lie just beyond the rapids. So, when the rains began the expedition once again set out. They were able to pass the first two rapids, but then they found they could go no further, and Livingstone anchored the boat. He, Kirk, and a few other crew members scaled the steep cliffs to assess what lay beyond. What they discovered was that the utterly impassable barriers continued for as far as they could see. This ended all speculation, even for Livingstone, about the feasibility of sailing up the Zambezi into the interior.[246] They would look for another way.

Continued failure and recall

The Shire River, which flows into the Zambezi, marked the location of the expedition's next attempt to reach farther into the interior. It was Christmas Day, 1858, and with the exception of Kirk and Livingstone all of the British members of the party were ill. Kirk and Livingstone climbed to the top of Mount Morrumbala, a 4,000-foot peak near the confluence of the two rivers, to assess the navigability of the Shire. From this point it appeared to be as wide as the Zambezi, and the recent rains had made the channel more than deep enough for the *Ma Robert*. The expedition immediately began following the Shire upstream, but only six days later the deep and smooth-flowing river entered into a vast stagnant swamp. The *Ma Robert* eventually forced its way through the sluggish waters and again entered the river's relatively straight channel, where it soon came across a village. The chief welcomed the travelers and told them of a great lake located north of his village. This lake, he told them, was the source of the Shire River. He also warned them of rough water between his village and the lake and said that he felt the *Ma Robert* would be unable to pass through it and reach their destination.

After finding that the Zambezi would not be the hoped-for "God's highway" into the interior, Livingstone needed a new claim to fame, and disregarding the chief's warning he insisted that they proceed at once. But the river soon narrowed, and the rapidly moving water tumbled through a series of rapids that extended upriver for over 30 miles. When the *Ma Robert* could proceed no further, he ordered the ship to return to Tete.

Defeated by both the Zambezi and Shire rivers, Livingstone desperately needed a victory. He decided that he and Kirk would take a few men and travel by small boat. They would leave the river to bypass the rapids, and then rejoin the Shire farther upstream. He could then lay claim to being the first European to see the source of the Shire and make the discovery of one of the largest lakes in Africa.

Unbeknownst, however, to the other members of the expedition, Livingstone already knew that he was not the first white man to see the lake, as the Portuguese trader, Candido José da Costa Cardoso, had reached it in 1846. This was not an ancient legend that might or might not have been true—Cardoso was still alive, in good health, and living in Tete. He and Livingstone had spoken at length, and the trader had even drawn him a map showing his journey to, and around, the lake. This did not, however, stop Livingstone from taking complete credit for the discovery. In his *Narrative* (more about this below) he states, "We discovered (the lake) a little before noon of the 16th of September 1859." He named his "discovery" Lake Nyassa (*sic*).[247]

The group had now been in Africa for nearly two years, and the men's contracts would soon expire. Little progress had been made toward the stated goals of the mission, the two rivers had proven impassable, and the British members of the crew were openly discussing not extending their contracts for another term. Kirk and most of the others were suffering from various illnesses including malaria, dengue fever, and dysentery. Livingstone's behavior was becoming more and more erratic. In addition to illness and the unpredictable behavior of the expedition's leader, tensions were now being created by the uncertainty of the supply lines. The supply ships from England drew too much water to enter the Zambezi. This meant that at prearranged times the *Ma Robert* would leave Tete to make the 11-day trip to the mouth of the river to collect supplies. Sometimes the supply ship would be there. More often it would be delayed by several days, or miss the supply drop completely. The expedition members had been promised that ships would be sent from England on a regular basis; food and essential equipment needed to survive, along with communication to and from the outside world, had been guaranteed. In addition to items flowing toward the expedition, a rendezvous with these transports also gave Kirk the essential mechanism to send his botanical collections back to Kew Gardens.

The strain was impacting everyone. Kirk, although not confronting Livingstone directly, found his moodiness and increasingly erratic behavior difficult to tolerate, and by January 1860 he too was planning to leave at the end of his contract. Livingstone was aware that members of the expedition were unhappy with his leadership. The dismissal of Captain Bedingfeld had been followed by similar actions against Baines and Thornton. Baines had been accused of stealing, and Thornton had been charged with dereliction of his duties related to searching for mineral deposits. All three men were well connected within the government circles. Livingstone knew that the men in London who would determine his fate already knew of the lack of progress and his unpredictable behavior, and that he had fired three key members of the expedition. He was fully expecting word from England that he was being recalled.

Therefore, he was both surprised and relieved when on November 23, 1860, a supply transport delivered a letter from London. He had earlier sent a letter to the head of the Foreign Office touting the promise of a third river, the Rovuma, as the probable highway into the interior. It transpired that based on his optimistic proposal the government was not only renewing the expedition's contract but shipping a new steam launch to them. In addition, a group of missionaries was due to arrive shortly and would take up residence in the areas that Livingstone had suggested. The goals of the mission seemed suddenly on the threshold of success.

But in February, the confident and positive mood of the expedition once again began to fade. On the way downriver to meet the missionaries and collect the new launch, the *Ma Robert* sank. All hands were saved, and the group proceeded in borrowed canoes to find, as planned upon reaching the ocean, Bishop Charles Mackenzie and four missionaries awaiting them. They also saw their new steam launch, the *Pioneer*, being towed by one of the British ships. Livingstone, eager to see how the launch would perform under the command of the newly assigned Captain May, convinced May to transport the group to the mouth of the Rovuma. But Livingstone's careless eagerness was not rewarded, as the *Pioneer* drew too much water to travel beyond the delta. His next decision was to take it out to the Comoro Islands to buy meat for the crew.

A subsequent argument between May and Livingstone over command of the vessel resulted in May resigning and Livingstone assuming command of the *Pioneer*, as he had done with the *Ma Robert*. Finally, the extended sea voyage, followed by travel up the Zambezi and Shire rivers, took a toll on the newly arrived missionaries. Mackenzie and one of the missionaries contracted malaria, and although they were able to recover under Kirk's care and the use of a quinine treatment, they had seen enough to start questioning Livingstone's leadership and judgement.

Bad luck continued. In January 1862 Mackenzie's sister and David Livingstone's wife, Mary, arrived, along with a new group of missionaries. With them was another steam launch, the *Lady Nyassa*, which Livingstone had personally paid for with the profits from his book. The sections of the new ship were placed onboard the *Pioneer*, which then headed upstream to join Mackenzie and the other missionaries. But when they reached the rendezvous point they learned that both Mackenzie and one of the original missionaries were dead. Kirk, on asking about the two men's symptoms, believed they had most likely died of cholera. Mary Livingstone began showing signs of illness soon after her arrival, and she too quickly passed away. In this case, however, Kirk attributed her death to malaria. Shortly after the passing of Mary and the two missionaries Kirk confided to his journal, "I can come to no other conclusion than that Dr Livingstone is out of his mind and a most unsafe leader." [248]

Regardless of his personal and professional losses, Livingstone was still determined to sail up the Rovuma. Despite the objections of almost the entire crew, it was decided that they would enter the river on September 9, 1862. Livingstone, now in one of his manic phases, was certain they could travel on the river to Lake Nyasa, but after days of agonizingly slow progress they encountered an Arab traveler who told them that the river would soon narrow down into a stream too small for even little boats to navigate, and that the river did not lead to Lake Nyasa.

Livingstone was devastated by this news, and at this point had no choice. He agreed to turn back. The Second Zambezi Expedition was, in every sense of the phrase, going nowhere.

The expedition was in a shambles—many of the original members had died or left the expedition, and almost all of those remaining were ill. After almost five years in East Africa they could point to few signs of success. They had charted Lake Nyasa and gathered information about the land and people of eastern and central Africa, and Kirk's observations and collecting had added to the body of knowledge regarding the animals and plants of the region. But beyond these meager accomplishments were nothing but failures. The explorers had failed in their attempts to navigate the Zambezi, Shire, and Rovuma rivers. The slave trade hadn't been stopped. Far from it in fact, the number of enslaved people being taken to Quelimane and Zanzibar had actually increased during the expedition's time in Africa, and the routes into the interior that had been opened by the expedition were now openly being used by the Arab slave traders. The string of missions had also failed to materialize. While Livingstone had scattered the missionaries in villages throughout the region, entirely isolated, and unable to receive supplies on a regular basis, they had been forced to scavenge for food and shelter. The few missionaries still alive in those remote outposts were besieged by disease and by hostile natives who wanted neither to give up slavery nor to accept the Christian god.

Reports of illness, death, and continual disagreements among the expedition's members had reached London. Livingstone's grandiose promises of vast mineral wealth and large fields of cotton had not materialized, and his increasingly unpredictable and erratic behavior was now well documented. Slavery was flourishing, and the missions were almost nonexistent. Therefore, he was not surprised when a letter from the Foreign Office arrived, saying:

> Her Majesty's Government fully appreciate the zeal and perseverance with which you have applied yourself to the discharge of duties entrusted to you. They are aware of the difficulties which you must necessarily have met with and they have deeply regretted that your anxieties have been aggravated by severe domestic affliction. Her Majesty's Government cannot, however, conceal from themselves that the results to which they had looked from the expedition under your superintendence have not been realized.[249]

It is clear that in the eyes of the British Government the expedition had been a complete, and costly, failure. A second letter stated that a ship would arrive in August 1863 to take the remaining expedition members back to England. The *Pioneer* was to be returned to the Royal Navy. Livingstone attempted to sell the *Lady Nyassa*, first in Zanzibar and then in Bombay. But finding no interest at either stop, he sailed it back to England, accompanied by only a few of the African porters.

Zanzibar

Livingstone was welcomed back upon his return to London, but not with the same degree of enthusiasm he had experienced following the First Zambezi Expedition. Members of the team and returning missionaries had been open about expressing their problems with his exaggerated claims and erratic leadership. The missionary community was especially critical, as they felt that his false assertions of a healthy climate and people eager to accept Christianity had led directly to the deaths of many who had taken him at his word. The Second Zambezi Expedition was castigated as a failure in many major newspapers. But to the general population Livingstone had become an almost mystical figure, and with the support of the public he felt undeterred. He quickly wrote his *Narrative of an Expedition to the Zambesi and its Tributaries; and of the Discovery of Lakes Shirwa and Nyassa*. His latest book became another top-selling publication.

Livingstone had already decided he would go back to Africa. But this time he experienced difficulty in raising funds for further exploration, as neither the missionary community nor the government was interested in financially supporting his exploration efforts. Instead, he would use the profits from his book to search for the source of each of Africa's two greatest rivers—the Congo and the Nile.

Kirk was also well received upon his return to England. He initially worked with William Hooker at the Royal Botanical Gardens Kew, and the abundance of plants he collected formed the basis for the *Flora of Tropical Africa*. Kirk consulted with Livingstone on *Narrative of an Expedition to the Zambesi*, and even offered some of his photos to be included, but he avoided any semblance of being a co-author. While others had been openly critical of Livingstone, Kirk had continued to limit any adverse comments to his private journal. Livingstone appreciated what he assumed was steadfast loyalty and invited Kirk to join him on the next expedition, but Kirk refused. He explained to Livingstone that he was declining to join it because he was about to marry and needed a stable job and sufficient income to support his wife, Helen Cook, and their anticipated family. Kirk had told her that they would not marry until he was assured of a decent salary, and she had agreed to defer the date.

He assured Livingstone that he wanted to return to Africa, but wanted options that would support him financially and in a location where he could be joined by his new wife. Livingstone then suggested a position that could address all of Kirk's needs.

Zanzibar was now under the direction of a friend of Livingstone's, Sir Henry Bartle Frere, the Governor of Bombay. So, Livingstone recommended Kirk to him for the consul position. A decade in the future Frere and Kirk would work together as they attempted to end slavery in East Africa. But for the consul position Frere was looking for a more experienced individual, and Kirk was passed over for the post. Instead, he was offered the vacant position of surgeon to the British Consulate. He accepted, but felt the salary of a medical officer would still not provide enough income for him to marry. He again turned to Livingstone, who was able to convince Sir Henry to add two additional positions to Kirk's duties: assistant to the political advisor, and vice-consul. These additional responsibilities would add significantly to his salary. Helen urged Kirk to accept the newly combined positions, and he did.

He left England in the spring of 1866 and at the beginning of June he arrived at his new home, Stone Town. This thriving community had begun as a small Swahili fishing village. But from the early 16th century the Portuguese had worked to develop its natural harbor into a major trading center. The Portuguese had ruled Zanzibar for over two centuries, developing its first stone structures, including the fort, which dominated the harbor. Toward the end of the 17th century the Sultan of Oman had been invited by the indigenous local citizens to overthrow the Portuguese. This having been accomplished, the sultan's descendants still controlled Zanzibar at the time of Kirk's arrival.

Although the Stone Town of Kirk's time contained 80,000 people, only 66 men, women and children were Europeans. Of this number, 22 were British. Beginning in about 1830 the European community, along with wealthy Arab and Indian merchants, called Banyans, began building additional stone houses near the harbor. This was where the consulates belonging to the British, French, United States, and Germany were located. The name "Stone Town" refers to the use of coral stone as the main construction material, giving the entire town a reddish color.[250] The heart of the city consisted of a maze of narrow alleys lined by the stone structures, interspersed with bazaars and mosques. The closeness of the streets and houses, the lack of a drainage system and sanitary regulations, and a drinking water supply that was utterly impure, plus the hot and humid climate, produced regular outbreaks of malaria, cholera, and other dangerous diseases and infections among the population.

The island's commerce was controlled by a very small percentage of the population, which was comprised of approximately 4,000 Arab plantation owners and 6,000 Banyan merchants. The vast wealth being collected from the island and from the African mainland only 30 miles away allowed these two groups to flourish through their trade in cloves, spices, and ivory. But the major commodity passing through Stone Town was enslaved people. Across the entire island of Zanzibar approximately 300,000 people resided. Within this total population, 220,000 were Africans, most of whom had been enslaved. It is estimated that between 500 and 2,000 enslaved people per month were sold in the markets of Zanzibar. Kirk, a lifelong opponent of slavery, had now found himself at the very center of the East African slave trade.

Following a six-month period of settling in, he wrote to Helen saying that he was ready for marriage. But his letters also expressed his anxiety regarding what he was asking her to do and the risks he was asking her to take. He warned her about the issues of disease, the lack of a European community, the obstacles he would encounter as he worked to end the slave trade, and the scope of work related to his three positions, which he would be expected to address on a daily basis. He advised her to think carefully about joining him. Six weeks later he received her response; she was packing and would soon be on her way. She arrived in Zanzibar on March 14, 1867, and they were immediately married in a Church of England ceremony aboard a Royal Navy ship in Zanzibar Harbor.

Prior to her marriage Helen had never traveled outside England, yet she soon created a comfortable home for the two of them. She also functioned as Kirk's secretary until the British Government assigned someone in that role. As the Kirk family grew (they eventually had five daughters and a son) Helen continued to balance her work, assisting her husband along with raising their large family. By 1875, in addition to their quarters in Stone Town, they had also purchased a home on the coast a few miles out of the city. This allowed Kirk to create a large botanical garden where he could cultivate an abundant plant collection from specimens he had acquired on searching throughout Zanzibar in its hot, humid climate. Shortly after his arrival, he had written to Helen that they would "not be bound to Zanzibar. We shall see, and stay or go, just as we find best." The two of them would spend 20 years together on this island home.[251]

Kirk continued during his first few years to function in the three positions Livingstone had negotiated, including surgeon to the British Consulate. As a result of that position he was soon spending large amounts of his time attending to the medical needs of the personnel connected to the German, French, and United States consulates as well. His medical reputation had also caused him to be selected by the sultan as the ruling family's personal physician. Yet, in spite of this hectic schedule, Kirk continued to expand his knowledge of languages. Within a few years of his arrival he was fluent in Spanish, Portuguese, and French, which served him well in his diplomatic duties. More importantly, he also became proficient in Swahili and Arabic; his competence in Swahili allowed him to collect information from informal sources within the general population of the city, and his proficiency in Arabic would be pivotal in his future negotiations with the Arab rulers.

Livingstone, Stanley, and Kirk

David Livingstone is often listed as the greatest of all the African explorers. Following the Zambezi expeditions, his books certainly made him the most well-known. He was undoubtedly a courageous explorer whose first-hand reports heightened awareness of the African slave trade.[252] But following the failure of his last expedition, things had changed. And, as mentioned earlier, he was forced to fund the next expedition almost entirely from the proceeds of his writing. On previous expeditions he had been accompanied by an extremely large and diverse group of fellow Britons. But now, seeking the sources of the Congo and Nile rivers, he would travel accompanied only by his African porters and two assistants, Abdullah Susi and James Chuma, who had been with him on the latest Zambezi expedition. Livingstone assigned these men significant roles within his search for the source of the two rivers. They became close aides, taking charge of the group when he became too ill to do so himself. After Livingstone's death they would play a pivotal role in his final journey.[253]

Although Kirk did not accompany Livingstone on this expedition, he still played an integral role. Livingstone's first stop was at Zanzibar, to collect the equipment and supplies needed for his journey, and to leave money with Kirk to pay for essential stores, then send them to pre-arranged locations for Livingstone to collect as needed. But this was problematic, for two reasons. First, Livingstone was looking for the starting point of the Congo and Nile rivers without the slightest idea of where either source was located. Unsure of where to begin, he was often known to travel in circles. Moorehead states, "Never can there have been a journey which was founded upon so many misassumptions as this one."[254] The reality was that the great explorer David Livingstone would spend the remainder of his life searching for the sources of the two rivers in regions where they did not exist.

Second, coordinating the timing and location of when and where to deposit the supplies would prove to be a constant problem for Kirk. Because the provisions had to be easy to find, they were also easy to steal, and much of the medicine and food was stolen from the rendezvous locations by those charged with taking supplies to the interior or by slavers. Too often supplies never reached the expedition. Livingstone faced constant struggles, in that he was attempting to abolish slavery while alone, unarmed, unsupported, and with no power to enforce existing laws. Yet he managed to struggle on, and in November 1871 he was in the village of Ujiji.*

For years rumors of Livingstone's death had been reported in England. The first report had been circulated by the porters who had deserted him early in the journey. Probably to cover their theft of medicine and supplies, upon their arrival in Zanzibar, they claimed they had returned because Livingstone had been killed on the shores of Lake Nyasa. Alternating reports of his being sighted in some remote location would be followed by yet another report of his death. But finally, letters from Livingstone to Kirk verified that the explorer was alive, even though his exact location was still unknown. It would take one of the most controversial figures in African history to ultimately locate the intrepid explorer.

Henry Morton Stanley, an American, had been born John Rowlands, a Welshman. His mother was an unmarried servant, and John would be the first of her three children, each having a different father. After very difficult early years, much of the time spent in orphanages and workhouses, Stanley traveled to the United States. Now in a new country, unknown to anyone, he gave himself a new name and a new identity. Stanley tells the story of being adopted by Henry Hope Stanley. However, according to Bierman, "not one scrap of documentary evidence [for the adoption] can be found." [255]

Whatever the story, around 1860 John Rowlands ceased to exist and Henry Stanley emerged. Living in New Orleans at the outbreak of the Civil War, he first joined the Confederate Army. He was captured following the Battle of Shiloh and, given the option of waiting out the war in a prison camp or changing sides, he promptly joined the Union Army. Following the war he became a journalist and served as a special correspondent for the *New York Herald*. At that point the whereabouts of Livingstone and determining if he was alive or dead was the top story throughout the world, and the paper's editor, James Gordon Bennett, sent Stanley to Africa to solve the mystery.

Stanley's and Livingstone's eventual meeting, along with the remarkable story of Susi's and Chuma's exploits following Livingstone's death, are perhaps the most frequently told stories in African exploration history. On November 10, 1871, Stanley and his caravan marched into Ujiji. Walking toward a group of Arabs, Stanley sighted a pale, weary man with white whiskers and mustache: "Dr. Livingstone, I presume?" [256] Although Stanley had been directed to find Livingstone or to verify the date and the cause of his death, once he had found him quite ill but very much alive the option of returning with him to Zanzibar never seems to have been considered. Instead, Stanley gave Livingstone as much of his own medicine and food supplies as he could spare, then traveled on to Zanzibar, reaching it in only 54 days. In August 1872—nine months later—porters and supplies sent by Stanley arrived at Ujiji, and Livingstone promptly left the village to continue his explorations.

Stanley had visited Kirk before leaving to begin his search, and now spent some additional time with him reporting his interactions with Livingstone. But although Stanley had the hero of his story, it now seemed that he needed a villain. He initially described Kirk as

---

* Halfway down the eastern shore of Lake Tanganyika, the border of what is now Tanzania.

cooperative and helpful, but then accused him of "letting Livingstone down" through his failure to keep him fully supplied.[257] This was repeated in his dispatches and in his numerous speaking engagements upon his return to England. Had Stanley's accusations and criticisms been proven truthful, Kirk's governmental career would have ended. However, two events quickly turned the situation around. First was the public response that Stanley received when he began voicing his allegations against Kirk as part of his lectures; the audience's' hostile response to them affirmed that Kirk was highly regarded throughout Britain. Second, one of Livingstone's letters reached the Foreign Office as Stanley initiated his charges. Livingstone's correspondence stated, "I never had any difference with (Kirk), though we were together for years, and I had no intention to give offence now." This would seem to have cleared Kirk of any wrongdoing, but the newspapers would not let the story die. This prompted the Foreign Office to send a commission of inquiry to Zanzibar to consider the accusations. The commission spent weeks investigating the claims, and ultimately its report gave Kirk its complete support. He was exonerated on all points, and Stanley was forced to retract his allegations.[258]

Now that Livingstone had been found alive the public's fascination with him seemed to evaporate. But the great explorer was still looking for the sources of the Nile and Congo rivers, and his health was continuing to fail. By the spring of 1873 he was wandering around Lake Bangweulu in present-day Zambia and was once again forced to depend on the support of Arab slavers to survive. He had become so weak that he had to be transported to Ulala, a small village, by litter.

On the morning of May 1, 1873, Susi and Chuma entered Livingstone's hut and found him dead. They removed his heart and buried it near where he had died. They then dried the body in the sun for a fortnight. It was wrapped in calico and placed in a bark cylinder which was wrapped in sailcloth and tied to a pole that allowed Susi and Chuma to carry it between them. They set out for Zanzibar, nearly 1,600 miles away, and the journey would take them over 11 months. In February they arrived in Bagamoyo on the Indian Ocean, from where HMS *Vulture* took Susi, Chuma, and Livingstone's remains to Zanzibar. From there, still in the company of his faithful servants, Livingstone was transported by ship to Southampton, and then by special train to London. On April 19, 1874, Livingstone's remains were interred in Westminster Abbey, with Stanley and Kirk serving as pallbearers.[259]

Churchill, Barghash, and Kirk

Zanzibar had been under the authority of the Sultanate of Oman since 1698, when the Portuguese had been driven off the island. The first anti-slavery treaty impacting Zanzibar had been negotiated by a Royal Navy officer, Captain Sir Fairfax Moresby, in 1822. The purpose of the Moresby Treaty was to prevent the importation of enslaved people from lands ruled by the Omani Arabs into holdings governed by Britain. The treaty created what was referred to as the Moresby Line, which ran from the southernmost point of Said bin Sultan Al Busaidi's African territory through the Indian Ocean to the city of Diu on the north-west coast of India; west of the line the transportation of enslaved people was permitted, whereas east of the line it was prohibited.

Confusion arose from the start. The English version of the treaty placed the responsibility for stopping and searching vessels on the Omanis, and the Arab text placed the responsibility on the British. As a result, the slave trade continued essentially unabated for the next 23 years.

By 1840 Zanzibar had grown from an insignificant little town to the principal port on the western shores of the Indian Ocean; the combination of the island's trade in cloves and the mainland trade in ivory and enslaved people had increased the sultan's revenue tenfold. In 1845 a new treaty was negotiated by Captain Atkins Hamerton on behalf of the British. Articles within this agreement clarified expectations related to searching and confiscating slave ships and their cargoes. The primary issue of slavery was, however, not the transportation of enslaved people, but the forcing of free people into slavery. Yet this new agreement still allowed the capture and enslavement of the local populations throughout the East African kingdom controlled by the ruling sultan without limits or penalties. The earlier treaties in 1822 and 1845 had been followed by additional agreements in 1862 and 1864.

Henry Adrian Churchill had become Zanzibar's chief consul in 1865, and since his arrival had worked on anti-slavery issues. None of the previous treaties had been intended to eradicate the slave trade but had only limited transportation to the countries to which the enslaved people could be sent. Churchill saw the earlier treaties as a starting point for ongoing negotiations that could end the practice of slavery throughout East Africa. On October 7, 1870, Barghash bin Said became the new Sultan of Zanzibar. When Barghash assumed the throne, Churchill felt his first responsibility was to reconfirm the existing agreements with the new ruler. When the Consul asked Barghash to verify the anti-slavery agreements he heatedly refused. Churchill was astounded that treaties which had been in existence for nearly 50 years could be rejected out of hand—this was not the type of diplomacy he was used to. He then suggested that the two of them formulate a new treaty. Barghash's response was even more direct.

> You propose that we should send someone to you to discuss with you the terms of a new treaty; but it is unnecessary to discuss [a new treaty or] the one actually in existence, and the trouble resulting therefrom is quite enough for us and more than enough.[260]

Churchill felt that to give in on the first point of contention with the new sultan would place Great Britain in a position of weakness. This could lead to challenges to British authority from Germany or France, which also had a strong presence on the island. Feeling that a show of strength was called for, he elected to use Britain's ultimate force.

Through diplomatic correspondence Churchill threatened retaliation by the Royal Navy or Britain's complete withdrawal from Zanzibar unless the treaty was signed at once. Barghash delayed his response, looking for another approach. Possibly aware that Churchill's greatest fear was that of Britain losing control to another European power, Barghash skillfully turned to the Germans for protection. The exact nature of the communication between Barghash and Theodore Schultz, the German consul, may never be known. But what is clear is that the letter sent by Schultz to Churchill stated that if the conflict between the British and Barghash continued the Germans would back the sultan. By this point Churchill, struggling with both the negotiations and his personal health, realized that he was physically unable to enter into what he anticipated would be a long-term process. He knew, too, that failure could lead not just to the continuation of the slave trade but to the loss of Zanzibar. Now, with the German intrusion there was also the possibility of war, and he asked Kirk to take over the discussions.

For Kirk, Churchill's request could not have come at a worse time. Three times since the beginning of the century East Africa had experienced a cholera epidemic. But the epidemic of 1869–70 was by far the worst. Cholera had spread from the Arab communities on Zanzibar and along the coast far into the African interior. It had moved from region to region, and from village to village. By the end of the first year the death toll throughout East Africa had risen to over 400 per day. By the end of 1870 it is estimated that 12,000 to 15,000 residents had died within Stone Town, and 25,000 to 30,000 on the island as a whole. In addition to his consular duties, and keeping the Livingstone expedition supplied, the management of the epidemic also fell to Kirk. By all accounts he was successful under very difficult circumstances, yet amazingly only one of the Europeans within the city contracted the disease. It was one of Kirk's daughters. Although she survived, the personal impact of the epidemic could only have added to his consternation.[261]

In spite of this immense workload and the emotional issues surrounding his daughter's illness, Kirk apparently accepted Churchill's request without hesitation. His earlier interactions with Barghash and other members of the royal family had been positive, and the sultan was already well on his way to considering Kirk a trusted friend. In their first meeting since assuming control of the negotiations, Kirk was careful to represent himself as having the sultan's best interests in mind. He assured Barghash that it behooved him, both financially and politically, to reaffirm the existing treaties. Barghash's reaction to Kirk and his diplomacy was immediate; he relented, telling Kirk that he would honor the existing agreements.

Within a few weeks the relationship that Kirk had developed paid additional dividends. Barghash appointed Kirk as Deputy Judge for the Sultanate, placing him in charge of the trials involving illegal slavers. This remarkable decision by Barghash allowed him to appear to be complying with the British without incurring the wrath of the Arabs. Most remarkable of all, after first refusing to agree to existing treaties, Barghash told Kirk that he was willing to consider additional limitations on the slave trade. In a stunningly short period of time Kirk was able to secure the sultan's signature acknowledging the existing treaties and end his potential defection to the Germans.

But not all was well with Henry Churchill. His illness continued to worsen, and he was soon forced to return to England. So, in December 1870, having seen Zanzibar through the cholera epidemic and successfully steered Britain through the transition into Barghash's reign, Kirk was appointed Acting Consul for the Island of Zanzibar.

Ending slavery on Zanzibar

While the outcome of the Napoleonic Wars was still very much in doubt, the Omani Arabs maintained strong ties with both the French and British. But by 1812 it had become obvious that France was losing the war, so the Arabs placed their entire support behind the British. The price they had been forced to pay to align with Britain was the signing of an agreement restricting the slave trade.

This agreement, like the other early treaties, had minimal impact, as they only affected the slave trade at the point where a country took possession of this human cargo—in fact, despite the agreements the number of enslaved people traded each year had significantly increased over time. For centuries the Omani sultans had received a tax on every enslaved person shipped from their territories, and by the time Barghash assumed the throne this amounted to the Omani equivalent of over £100,000 per year.

Complete abolition of the slave trade would mean the loss of the biggest item in Zanzibar's customs revenue. It also meant that the ruler agreeing to these terms would be seen as the man who had ended a practice in which Arab society had been engaged for hundreds of years; his predecessors had simply signed the treaties, stopped shipping enslaved people to the European colonies in India, and turned a blind eye to the rest of the practice. Barghash saw no reason to change the routine that had continued through a succession of rulers.[262]

Even before he assumed the throne, the British were already planning ways to further restrict the slave trade. Shortly following Kirk's appointment as acting consul, the British Government appointed a Select Committee "to inquire into the whole question of the Slave Trade on the East Coast of Africa … and the possibility of putting an end entirely to the traffic in slaves by sea." Beginning in July 1871 the committee took evidence from 14 witnesses. From these witnesses and from Kirk's dispatches, the committee acknowledged the failure of the existing treaties and policies to restrict, or even address, slavery in East Africa. The strategy that emerged from the committee's work took a completely different direction from the strategy implemented in the past; the treaty, they now declared, must provide for the abolition of the slave trade, not its restriction. The sultan should be informed that if the trade was not stopped by other means the British Government would take the "requisite measures to put an end to all slave trade." [263]

Although Kirk was designated to serve the role of liaison between the committee and the sultan, it was not he who would lead the negotiations. That task would fall to Sir Henry Bartle Frere. According to Sir Reginald Coupland, the Beit Professor of Colonial History at the University of Oxford, "few … who had served the British Empire [had] acquired as high a reputation." Frere had already served as chief commissioner of Sindh, in present-day Pakistan, and for his actions he had been knighted and presented with honorary degrees from Oxford and Cambridge.

Frere (who as governor of Zanzibar back in 1863 had appointed Kirk to his positions at the Consulate), left England in November 1872. Upon arriving in Zanzibar in January 1873 he submitted the draft terms of the treaty to Barghash, who then met with his family members and advisors. They pleaded with Frere for a delay in signing the treaty, or for a modification of the terms to allow for a gradual abolition of the slave trade. Negotiations continued, three days passed, then Frere received his answer: "We cannot sign the new treaty." [264]

The negotiations were clearly at an impasse, and on February 15 Frere left Zanzibar to inspect the slave ports along the coast. Reports of what had happened and the rejection he had received preceded him, and as a result British prestige suddenly and sharply declined throughout the region. Frere returned to Zanzibar on March 12, 1873, and Kirk reported that the sultan's decision was unchanged. Frere, deciding to give Barghash one last chance, directed Kirk to convey that if the sultan had any counterproposals now was the time to submit them. Barghash's response was more emphatic than before, and because of these setbacks Frere decided to leave Zanzibar permanently.

Although he had failed to secure the treaty, he was not ready to give up and before leaving Zanzibar he presented Kirk with the outline he had created for the negotiations. The terms of the agreement were severe and left nothing to misinterpretation:

> First, the shipment of slaves brought from the interior of Africa should be regarded and treated as piracy. Second, the right conceded to Zanzibar subjects by implication in the treaty of 1845 to continue the shipping of slaves from port to port within the Zanzibar dominions should be withdrawn. Third, all slave markets should be closed. Fourth, an embargo should be placed on all Zanzibar customs houses to prevent the passage of slaves. Fifth, the [British] naval squadron should be increased to fourteen ships.

Frere also sent a message to the Governor of Bombay, urging him to implement the commission's recommendation and immediately make Kirk the Consul of Zanzibar.[265]

Through a telegram sent by Frere on June 2, Kirk received additional advice:

> Inform the Sultan that Her Majesty's Government requires him to conclude the Treaty … and that the Sultan take effectual measures within all parts of his dominions to prevent and suppress the same … If the treaty is not accepted and signed by him before the arrival of Admiral Cumming … the British naval forces will proceed to blockade the island of Zanzibar.

Upon receiving the telegram, Kirk informed the German and American consuls and received their verbal support. But as the French were secretly urging the sultan not to sign the new treaty, the best that he could get from them was an agreement not to interfere.

The next day Kirk met with Barghash and asked that his four chief advisors be present as well. Kirk, however, was alone and he conducted the negotiations entirely in Arabic. He first read out the telegram and explained that a blockade meant that no ships, Arab or European, would be allowed to approach the island. All trade with the outside world would be stopped. Kirk then turned to the advisors and said he hoped that they would not misdirect the sultan. He indicated that he was putting the responsibility for this crucial decision on their shoulders; if the treaty remained unsigned the blame could not be shifted onto the sultan. He concluded his brief remarks by declaring that this was an ultimatum, and its rejection meant war. With that he withdrew.[266]

That evening, Kirk was summoned to the palace again. He was asked if he could modify the terms. He would not make concessions. Barghash then announced that he would go to England personally for negotiations. Kirk told him it would be a mistake for him to leave the kingdom at this pivotal time, and that he could expect no concessions in London. Kirk left the palace but was soon sent for again. Barghash then gave his response: "Now I understand, and you may consider the treaty signed."

The next morning, June 5, 1873, Barghash, true to his word, signed into law the Treaty between Her Majesty and the Sultan of Zanzibar for the Suppression of the Slave Trade, and promptly sent messengers to clear the slave market. When his subjects there attempted to demonstrate, he sent in a company of his troops, and the protest was quickly and peacefully put down. A proclamation was posted to the effect that this was the "final act," and that it had been agreed upon by the sultan's chief advisors.

Kirk's negotiations could not have been more skillfully conducted. It is clear that the sultan trusted him as he trusted no other foreigner. They regarded each other as friends, and the two of them would now be traveling to England.

England and beyond

After the treaty signing, Kirk received accolades from throughout Europe and America. Kirk spent the next twelve months overseeing the enforcement of the new treaty and conducting surveys regarding its effectiveness, and July 1874 he traveled to England on a much-needed and well-deserved leave with Helen and their children. But it would become a working holiday; in May 1875 Barghash, having been denied his visit to England during the negotiations, followed Kirk to London. The sultan and his large entourage were honored by Parliament for having signed the antislavery treaty, and Kirk, along with Bartle Frere, served as the delegation's hosts during most of their visit. By all accounts, Barghash's time in England was a great success. He was invited to Windsor Castle, where he met Queen Victoria, and the Prince of Wales gave him and his entourage tours of Birmingham, Manchester, and Liverpool. The tour paid even greater dividends for the sultan within the Arab world; it was clear that he had been greeted as an equal by the leaders of the world's most powerful empire, and his standing among his peers rose substantially.

Although the sultan's visit to England was highly successful, the situation in Zanzibar was deteriorating. Barghash's enforcement of the treaty through his control of the ports, and Britain's increased Royal Navy patrols, had almost completely stemmed the flow of enslaved people by sea. However, in its place the traffic by land had grown substantially, as enslaved people were now being marched from the interior to ports beyond the territories controlled by Barghash.

As the use of British ground troops was not feasible either logistically or financially, Kirk felt that the situation might be addressed more effectively through what he called indirect action. If the sultan could be persuaded to use his authority throughout his mainland territories as he had used it within the port cities, the slave trade could be fully eliminated. Lord Derby, the British Foreign Minister, accepted this suggestion, and upon their return to Zanzibar, Kirk asked Barghash to stop the entire trade and to fulfill the real intention of the treaty "with a single stroke of the pen." Kirk stated that the British Government was aware of Barghash's "good faith and earnest wish to do all that has been asked," and added that they would unconditionally support him in these efforts. After a brief deliberation, the sultan responded that he was willing to do all that he had been asked.[267]

Unlike naval action, which relied exclusively on British ships, enforcement of the latest proclamation would depend mainly on the sultan's Arab troops. Kirk requested Barghash's permission for Second Lieutenant Lloyd Mathews of HMS *London* to train the sultan's army, and his request was honored. Impressed by the sultan's decisive actions the Foreign Office also sent him a gift of 500 new Snider rifles. With the British weapons, and under Mathews' direction, the sultan's forces soon grew into a 1,300-man, well-trained army. Barghash rewarded Mathews by naming him a brigadier general, and Mathews became one of the sultan's closest friends.

From that point on, after years of missteps and failures, the Arab slave trade as an organized business ceased to exist. Kirk could now take pride in the fact that he had accomplished what no man before him had been able to do. He had negotiated the final terms of the treaty to end slavery in East Africa and put in place the mechanisms for its enforcement.[268]

Missed opportunities

By the 1880s Africa was well on the way to being consumed by Europe. In East Africa, Germany, a newcomer to colonization, was taking the lead, in that expeditions supported by the German Government were traveling throughout the mainland and securing treaties with numerous local leaders for trade and settlement. With these treaties in hand, Germany formed a protectorate from the eastern flanks of Mount Kilimanjaro to the Indian Ocean. Although the Germans had taken control of a large part of Barghash's kingdom, Kirk felt that if the British intervened immediately Barghash could still retain the remainder of his territory. So, he telegraphed the Foreign Office, saying:

> It would assist me much if I did know whether under any circumstances
> the British Government in case of any opportunity offering, could now
> consider acquisition or a protectorate of a district with a naval port. I
> mention this believing that Zanzibar must soon break up or pass bodily
> to Germany.

The policy of the Foreign Office at that time was one of careful balancing and neutrality, with the final object of not offending any of the European powers. Lord Salisbury, the prime minister, denied the request and ordered Kirk "not to permit any communications of a hostile tone" to be directed toward Germany by either his office or that of the sultan; Barghash must agree to all of Germany's terms.[269]

Kirk had foreseen the importance of saving as much of East Africa as possible from the Germans. In this he had Barghash's full support, the sultan even offering to put the remainder of his kingdom under British sovereignty. Kirk conveyed this offer to the Foreign Office, who again rejected it without an explanation. Although the Germans had attained all that they sought, they had done so over Kirk's adamant protests, and the German chancellor, Bismarck, and emperor, William II, were not about to let this pass; the British prime minister and the Foreign Office were inundated with correspondence outlining what the Germans described as Kirk's "obstructionist behavior."

In March 1887 Kirk received a letter from the Foreign Office stating that he was needed in London to discuss how to best handle the German situation. Kirk recognized that this was a pretext for his dismissal, and that his career as a colonial administrator was finished. He had served in Zanzibar for 20 years, and his support of Barghash, as well as his insistence on following a course that he felt was best for Britain, had been too public. Thus, upon his return to England he was not surprised to find himself without a government position. He officially retired from consular service within a year of his return, citing health reasons, despite the fact that he was only 55 years old and remained physically fit.[270]

Honors and Sevenoaks

On an earlier visit to England the Kirks had purchased a house at Sevenoaks in Kent. Now, permanently removed from Zanzibar, they moved into their new home. Although officially out of government service, Kirk continued to be consulted regarding African affairs, and even undertook some missions on behalf of the Foreign Office. In 1889 King Leopold of Belgium convened a multi-nation conference to discuss the complete abolition of the slave trade and the freedom of enslaved people already held in captivity. Kirk was chosen as Britain's plenipotentiary,[*] and he, with his knowledge of Africa and the slave trade, formulated the proposal that became the treaty agreed to by the 17 nations in attendance. After ratification, the delegates passed a resolution of appreciation for the manner in which Kirk had guided the process to its successful conclusion.

For the final 30 years of his life, in addition to his demanding schedule of meetings and an extraordinary volume of writing, John Kirk was content to work in his garden at Sevenoaks. In 1879 Queen Victoria had awarded him the order of St Michael and St George (CMG). This was followed in 1881 by Knight Commander (KCMG). Then in 1886 Kirk was made a Knight Grand Cross (GCMG), the highest rank of the order. Other honors included the Patron's Gold Medal of the Royal Geographic Society in 1882 "for his long continued and unremitting services to geography in Africa." For his work in the natural sciences, the Royal Society elected him a fellow in 1887.

From 1907 onward Kirk suffered from partial blindness, and in 1914 his wife Helen died. Yet neither the loss of his sight nor the loss of his spouse prevented him from pursuing his two great avocations, gardening and photography, which he pursued for the remainder of his life.

---

* A diplomat invested with the full power of independent action on behalf of its government

In sum, John Kirk, the son of a Scottish minister and trained as a physician, had succeeded where others had failed. With his work in Zanzibar he had ended slavery in East Africa. Through his leadership of the Brussels Anti-Slavery Conference he had extended that work throughout Africa as a whole.

Sir John Kirk died at Sevenoaks on January 15, 1922. He was 90 years old. Sir Henry Johnson, a fellow East African explorer and a friend of Kirk's, wrote a fitting eulogy:

> I have always thought Sir John, one of the greatest men produced by Great Britain during the nineteenth century, was never properly appreciated until long after his official career was closed … It was to him that all those interested in central African questions – especially regarding the slave trade – turned for advice … He has been my counsellor and friend, for whom admiration vies with affection and whose example it has been my greatest ambition to follow – the ablest, the most sympathetic, and the most modest of all men.[271]

8

John Rae (1813–1893)

Nothing burns like the cold. But only for a while.
Then it gets inside you and starts to fill you up,
and after a while you don't have the strength to fight it.

– **George R.R. Martin**

By the early part of the 18th century the British ships traveling to and from the Hudson's Bay Company (known as the Company) outposts in what was known as Rupert's Land were routinely stopping for food and water at the port of Stromness in the Orkney Islands. Located on the south-western point of Mainland, Stromness is the second-largest burgh of the 74 islands comprising the Orkney Archipelago. **(Figure 8.1)** The port faces the Bay of Ireland, and directly across the bay is the parish of Orphir. Within the parish, on a hill overlooking

Figure 8.1 Stromness (Orkney Library Archives)

the choppy waters of the bay, stands the Hall of Clestrain, and in this house on September 30, 1813, John Rae was born.

He was the sixth of nine children born to Margaret and John Rae. At the time of the young John's birth, his father was serving as a land agent for Sir William Honyman, Lord Armadale, one of the most powerful men in northern Scotland. Honyman had become a lawyer and politician, and during the course of his lifetime amassed extensive land holdings, including large parcels within Orkney. Rae's father was responsible for supervising the 300 tenant farmers who worked Honyman's land and assisted in managing his extensive herds of cattle and sheep. In 1819, in addition to his duties related to Honyman's land holdings, Rae's father began serving as an agent for the Company, recruiting Orkney men and sending them out on the ships bound for Canada.

John grew up pursuing the same activities as the sons of the tenant farmers his father supervised. He enjoyed hiking, fishing, hunting, and boating—pursuits which would serve him well in his not-too-distant future. John and his siblings received a private education. The governess first assigned to provide the basics soon gave way to a series of live-in tutors who expanded the children's knowledge in their advanced studies. Unlike his siblings, whose education ended with their tutorials, John asked for more to his education, and at age 16 he enrolled in the University of Edinburgh Medical School, where he studied for four years. Upon completing his studies within the university, he, like many students at that time, enrolled in the Royal College of Surgeons to prepare for his surgeon's license. In April 1833, at 19 years of age, he passed his written and oral examinations and qualified as a Licentiate of the Royal College.

Employment in the Company in any capacity offered more than twice the salary the men of Orkney could earn at home performing the same tasks. The Company needed carpenters, clerks, trappers, and hunters—and it also needed skilled doctors to tend the hundreds of men now employed within its various outposts and factories. Rae's father had previously secured clerical postings in the Company for William and Richard, two of John's older brothers. John had always wanted to share in their adventures, and now, with his medical license in hand, it was possible. In June 1833 John Rae left Stromness Harbor as the surgeon aboard the *Prince of Wales*, a supply ship headed out on its annual voyage to Hudson Bay.

Moose Factory and the long cold winter

In addition to massive amounts of equipment and supplies bound for the Canadian outposts, the *Prince of Wales* was transporting 31 new employees who had been recruited from Orkney by John's father. Early in the voyage typhoid fever broke out among those steerage passengers, and many became seriously ill. The newly licensed Rae spent two weeks, working day and night, tending the sick. Though physically pushed to his limits, he completed the treatment without the loss of a single man, and on September 7, 1833, the crowded ship arrived at its destination near the Company outpost known as Moose Factory. (**Figure 8.2**)

Moose Factory got its name from its island location near the mouth of the Moose River. Because supply ships could sail directly from England and anchor within a few miles of the

*Figure 8.2 Moose Factory
(Wikimedia Commons/PD)*

factory, it became one of the two main fur-trading posts for the Company, and the landing point for most supplies and new employees committed to it. For several days following their arrival, small boats traveled out to the *Prince of Wales*, offloading equipment and supplies from it, and transferring them to the island factory. From there, these goods would be distributed to the outlying posts to support the men responsible for collecting the furs. As the ship was emptied of men and supplies the returning small boats were packed with pelts which had been collected from remote stations. The pelts were then stowed in the hold of the *Prince of Wales* in preparation for the return voyage to England.

During his brief stay Rae must have impressed the chief factor at Moose Factory, as he was offered the position of post doctor. He was tempted but declined. He felt his promise to return to his family must be honored, and a little over two weeks after arriving, he returned to the *Prince of Wales* for the homeward voyage. Delays both in traveling to the factory and in the offloading of the supplies meant that they were beginning their return home almost three weeks later than planned, and this, coupled with an early start to that season's harsh Canadian winter, meant that upon entering Hudson Strait they faced a solid sheet of pack ice. At the time of Rae's voyage, the Company presented a bonus to captains and crews who were able to complete the journey out and back during the same season so, spurred on by this potentially significant windfall, the captain tried every possible maneuver he knew to continue the homebound voyage.[272] But after trying for several days to find a route through or around the formidable barrier the captain gave up, and the *Prince of Wales* turned back toward the Canadian mainland.

The ship now had ice covering every part of its deck and rigging. Near the bow the ice was over two feet thick, making the vessel extremely difficult to steer. The men needed to find somewhere to settle in for the winter and needed to find it quickly. As large as Moose Factory was, it could not provide the additional housing needed to support the ship's crew for the winter. So, the *Prince of Wales* headed south toward Charlton Island, where the ship's crew had wintered several seasons before after battling the same problems with the pack ice. But two years had passed since that earlier winter, and the Company post had been abandoned. Worse, upon beaching the *Prince of Wales*, it was noticed that most of the buildings had lost their windows and roofs. Rae and the crew first erected a tent using the ship's sails. This was not for their own comfort, but to protect the valuable cargo of pelts that would be expected when they arrived back in England. Only when this task was completed could they begin working to repair what would be their home for the next several months.

The captain of the *Prince of Wales* sent a small boat back to Moose Factory with two paying customers who had made the unfortunate choice to return to England on that particular ship but would now instead be spending the winter on the Moose River. After several weeks, the boat and crewmen returned with additional food to assist in supporting the stranded group through the winter. They also brought along extra blankets and warm clothing. While the majority of the crew settled in for what they correctly assumed would be a long hard winter, John Rae experienced the thrill of embarking on his first real quest.

Having spent the entirety of his short lifetime hunting, fishing, and hiking in the harsh Orkney landscape, Rae seemed to welcome this new adventure with what can only be described as pleasure. Writing in his journal he describes fine dry snow that he found perfect for snowshoeing, saying, "Personally, I enjoyed the situation immensely." He also liked hunting unfamiliar game, and his shooting proficiency helped greatly to supplement the meager food supplies. Again, he writes, "I found (the hunting) extremely attractive, a feeling which was not altogether shared by the older portion of our party." [273]

As the winter wore on the men faced temperatures well below freezing, heavy snowfall, and high winds. They were often restricted to their meager shelter for days at a time, making Rae's hunting expeditions sometimes impossible. But the most concerning problem the stranded men encountered was not a lack of food, although it was not excessive in any sense of the word. Nor was it the coldness of the Canadian winter, although their lodgings could only be described as minimal. The problem was a lack of anything other than meat and biscuits to eat, and soon the entire crew began showing signs of scurvy. The disease was not yet completely understood, but it was known that fresh vegetables and citrus fruits could prevent or cure the disease. The *Prince of Wales* had provided a supply of lemon juice that would have lasted well beyond the time needed to return to England. But for the crew, now stranded for the entire winter, the supply soon ran out.

Rae did his best to treat the sick and experimented with a variety of potential cures such as boiled spruce needles. But nothing he tried proved to be an effective treatment; 17 of the men became quite ill, and their condition soon became critical. First, they experienced a swelling of their arms and legs, which quickly became so severe that they had difficulty moving about. There followed a swelling and discoloring of the gums that made eating anything difficult. Soon their teeth became so loose that they fell from their gums. At the time, scurvy was often fatal, and before the end of the winter both the captain and first mate had died from the illness.

Nearing the end of the winter, Rae was again able to resume his hunting, and by chance and keen observation was able to solve the scurvy crisis that was affecting the crew. While traveling across country he observed small red patches in the snow. Initially he assumed it was blood from a wounded animal, but soon realized that he had crushed cranberries under his snowshoes, and the red patches were made by their juice seeping up through the snow. Rae proceeded to clear the snow from what he found to be an extensive field of wild cranberries. He then rounded up the men still suffering from scurvy and took them out during the warmest part of each day. Over the period of a week he encouraged them to eat as many of the cranberries as they could gather. This treatment, along with some soup that he made from buds of the vetch,* restored the health of the remaining 15 crew members. [274] At

---

* A herbaceous plant of the pea family

the end of the winter the stranded party dragged the beached ship back into the water and began the return voyage to Moose Factory.

Upon his arrival Rae learned that the chief factor had sent a letter to his family, via the overland route to Montreal. He had told them of the stranded crew's situation, and he had also told them of his offer to employ Rae as post doctor. In a return letter Rae's mother had encouraged him to take the position, at least on a trial basis. In correspondence between the factor and the Company's governor, Sir George Simpson, it had been suggested that Rae be offered a five-year contract as surgeon/clerk. Rae responded that while he would accept the position of surgeon for two years he had no interest at all in becoming a clerk. The chief factor agreed to these terms, and in a letter to Simpson stated, "He's a very attentive and pleasant young man, hardy and well-adapted to the country … and may in time take a notion of remaining."

Throughout the range of the territories controlled by the Company the living conditions were harsh. Although Moose Factory was one of the oldest and largest of the Company's outposts, conditions were still quite primitive, and few men ever extended their stay beyond their initial contract. Rae, having grown up in a rugged environment that was hundreds of miles farther north than the outpost, adapted better than most. His abilities as an outdoorsman meant he thrived in a winter climate that many of his fellow workers failed to appreciate. He also possessed what was described as an "egalitarian" outlook toward the native peoples. While other Company men attempted to stay away from local inhabitants for fear of "going native," Rae seemed to prefer the company of Cree and Inuit men.* He chose them as traveling partners, and he tried to learn all he could from them. His innate skills in the wilderness, his ability to live off the land and his pleasure in doing so, combined with a sincere admiration for the native people's culture and traditions, would make him unique among the Arctic explorers.[275]

Rae soon understood why his initial offer of employment had been a combined position of doctor and clerk; his employment as a doctor gave him too little to do. So, he began looking for other ways to occupy his time in the Company. He volunteered to work at everything from sorting supplies to processing the furs that arrived throughout the winter. He always took opportunities to travel by snowshoe, exploring wider and wider arcs around the outpost. He was a superb hunter, and when traveling by boat or canoe he had few peers amongst the Europeans. When traveling by snowshoe he was matched by none of them. Robert Campbell—a fellow Scotsman and employee of the Company, and a fur trader and explorer who established a distance record of 3,000 miles on snowshoes—stated, "(Rae) was the best and ablest snowshoe walker not only in the Hudson's Bay Territory, but also of the age." [276] Rae would spend the next 10 years at Moose Factory and over 20 years in the employ of the Company.[277]

The Search for the Northwest Passage

During the first half of the 19th century traveling from England to the major trading points in Asia such as those in India, China, or Japan was not an easy task. Before the Suez Canal

---

* Two of the larger First Nations groups living in Canada

opened in 1869 one option was to skirt the coast of West Africa and sail around the Cape of Good Hope. The second, prior to 1914 when the Panama Canal opened, was to cross the Atlantic Ocean and reach the Pacific by rounding Cape Horn at the tip of South America. These journeys were long and costly, and the seas around both capes were always dangerous. Ships encountered strong winds, large waves, and treacherous currents—and, when traveling around Cape Horn, icebergs. To this day, traveling around either cape is still regarded as one of the major challenges to shipping. Until near the end of the 19th century dozens of ships were lost in these regions each year. So, Europeans in general, and the British in particular, were desperate to find new options.

The ideal solution was the fabled Northwest Passage around North America. By 1842, about the time Rae was nearing the end of his ten-year stay at Moose Factory, Arctic exploration to find such a passage dominated discussions throughout the shipping industry in England. An early charter given to the Company had included a clause directing it to devote personnel, equipment, and finances to search for the passage. In England, merchants and members of Parliament were now becoming openly critical of the Company's lack of accomplishments to this directive. It needed to exhibit success, or at least a highly visible attempt to locate this long-sought-after channel. As Rae had been in the company's employ for ten years, his skills were well known. Sir George **(Figure 8.3)** had heard of Rae's accomplishments, and he decided he would transfer the responsibility for the search for the Northwest Passage to his young doctor.

Late in 1843, after swearing Rae to secrecy, Sir George invited him to spend Christmas with him in his Montreal mansion. While Rae was enjoying splendid gubernatorial hospitality, the two of them worked out the details of what would be the most complete Arctic survey to date. Rae would attempt to link the discoveries of Royal Navy explorers, Sir John Ross and Sir Edward Parry, who had traveled from the east in the 1820s and 1830s, with those of Company employees Thomas Simpson and Peter Warren Dease, who in 1839 had searched for the passage from the west. Rae would attempt to locate a passage that would link these two previous efforts, finally proving the existence of the elusive Northwest Passage. Following his stay of over a month in Montreal, Rae returned to Moose Factory, traveling most of the 700 miles on snowshoes and hauling a loaded sled the entire way. Later, in May, Sir George sent a private letter asking to meet with Rae at Moose Factory to discuss a private matter.

> An idea has entered my mind [he wrote, as if they had never met at Christmas,] that you are one of the fittest men in the country to conduct an expedition for the purpose of completing the survey of the northern coast that remains untraced … As regards the management of the people and endurance of toil, I think you are better adapted for this work than most of the gentlemen with whom I am acquainted in the country, and with a little practice in taking observations, which might very soon be acquired, I think you would be quite equal to the scientific part of the duty.[278]

Figure 8.3 George Simpson (Wikimedia Commons/PD)

Rae's first challenge, then, had nothing to do with the elements or the rugged terrain that he would encounter; the directive was not simply to discover the Northwest Passage but to survey, map, and chart the correct route—and Rae knew nothing about any of these disciplines. For Rae to acquire these skills, Simpson directed him to spend the winter of 1844–45 at the Company's Red River Settlement, where his designated instructor would tutor him not only in the techniques needed in surveying, but also in other sciences such as astronomy, geology, and botany. After hiking for two months from Moose Factory to Red River, Rae arrived at the settlement, only to discover that the man to whom he had been assigned was very ill. Two months later his anticipated tutor died. Rae was now faced with the prospect of either setting out without any training or delaying the exploring journey until he could find another instructor.

After consultation with those familiar with the needed credentials, Rae concluded that the most suitable replacement for teaching him the skills needed for his journey was located at Sault Ste. Marie in northern Ontario. In January, along with three other men, he set out on the 1,200-mile trek. The journey took almost two months. But when Rae arrived, he discovered to his dismay that his second likely instructor was no longer stationed at this border town.

Now unclear on what his next step should be, he sent a message to Simpson asking for advice. The letter and its response took another four months, and Simpson directed him to continue to Toronto. Here Rae would be working at the Toronto Magnetic and Meteorological Observatory with Captain L.H. Lefroy, the leading astronomer and surveyor in Canada. While in Toronto Rae received the official resolution from the Company, authorizing his expedition. He learned he would command two small boats containing 12 men. He was to sail north from Churchill, located on Hudson Bay above the York Factory, the largest facility within the Company, and chart the Arctic coastline to the mouth of the Castor and Pollux River.

After completing his training with Captain Lefroy, Rae traveled to the York Factory, arriving in October 1845. He initially experienced difficulty in securing men he felt were up to the challenge and it was not until June 1846 that the ten men that would be joining him were fit, motivated, and ready to depart. Although locating a suitable crew had produced yet another delay, in the end Rae was satisfied with the results. Six of the party were Scots: four from Orkney, one from Shetland, and one from the northern part of the mainland. Two were French Canadian, one was Métis,[*] and one was Cree. Upon leaving the York Factory, the assembled group pushed north for two weeks, and the men's consistent fast pace and stamina reinforced Rae's confidence in them. Making remarkable time, they soon arrived at Fort Churchill, where Rae added two Inuit hunters to the group.

The next morning, using the two small boats constructed for the expedition, the party sailed north and began exploring, charting, and surveying the shore of Hudson Bay, passing Repulse Bay.[†] Rae fished, and shot geese and ducks, to supplement and extend their food supply. He also collected specimens of birds and plants to be shipped to England upon his return. But in August the party began to encounter sheets of ice, and it soon became apparent that with winter setting in they could travel no farther north. They returned to Repulse Bay, and prepared to meet a challenge that no European had yet met; they would spend the winter in the High Arctic, relying exclusively on their own resources.

By early September they had constructed a stone house using hides for the roof, door, and window coverings. Game was plentiful, and the party relied almost exclusively on Rae to provide the needed provisions. Rae and the European members of the crew also observed the Inuit building their igloos. After a few failed attempts, some of the party became adept at ice construction, and several igloos, connected by tunnels, were built around the original stone structure, and were used for storing supplies. During that winter the temperature reached a chilling low of –47ºF–44ºC, yet through careful planning, skilled hunting, and certainly good instinct, the group survived the treacherous winter.

Meanwhile Rae, undeterred, was already planning for their departure in the spring. In spite of the consistently cold weather, he directed two of the party to begin construction of sleds which would be used for the next portion of the expedition. In April, he set out with five of his men to further explore the region. They traveled almost non-stop for 30 days, covering nearly 600 miles. Upon their return they rested briefly, then set off to chart the western coastline of the Melville Peninsula. Rae and his party eventually arrived back at Fort Churchill on August 12, 1847.

The journey ultimately had a mixed outcome. His party had survived the winter in the High Arctic, feeding and sheltering themselves without outside assistance, and although he had surveyed almost 700 miles of new territory, accomplishing all this without losing a single man, meaning that his methods would be replicated by Arctic explorers who followed him, the goal of his exploration had not been achieved. He had yet to find the missing link for the Northwest Passage.

The lost Franklin Expedition

Although the Hudson's Bay Company had been designated to search for the Northwest Passage, the British Government apparently did not feel that a joint-stock company could unilaterally accomplish the undertaking. In May 1845, before Rae's group set out from the

_____

[*] A person of mixed indigenous and Euro-American ancestry

[†] Now Naujaat

York Factory, the Admiralty had mounted a naval expedition to search for the Northwest Passage; this was the same geographic riddle which John Rae had been assigned to solve. A list of over a dozen candidates to lead the voyage was considered, and from this group the Admiralty selected Sir John Franklin.

Franklin, although a veteran Royal Navy officer with three previous Arctic explorations, was not an obvious choice. He had first visited the Arctic on sailing from London in 1818 with David Buchan. Each man had commanded a converted whaling ship outfitted with reinforced hulls. Their charge had been to discover the Northwest Passage by forcing their way through the 2–3-meter-thick ice of the Arctic Ocean. After nearly losing both ships in the packed ice, and taking three weeks to extricate themselves, they returned to England to report that there was no Northwest Passage.

Then between 1819 and 1822 Franklin led the disastrous Coppermine Expedition. His orders were to travel overland from Hudson Bay and chart the coast of Canada from the mouth of the Coppermine River. His party included John Richardson as surgeon and naturalist. Even though Franklin's voyage produced little new information, and he lost 11 of his 20 men, most through starvation or exhaustion, his account of the rather unproductive expedition became a best-selling book. Successful or not, Franklin became permanently linked to Arctic exploration.

In 1826 Franklin and Richardson returned to map the coast between the Mackenzie and Coppermine rivers. This expedition proved much more successful, and upon his return from this third effort Franklin was knighted and then named Lieutenant Governor of Van Diemen's Land.* However, in 1843 he was recalled from his post after being censured for incompetence. Nearing 60 years of age, he wanted one last opportunity to salvage a fading reputation. Both he and his wife, the socially and politically well-connected Lady Jane Franklin, lobbied strongly for him to be selected to head this new expedition to find the Northwest Passage. In what came as a surprise to many, he was chosen to lead the search.

Toward the end of July 1845, while the expedition's two ships, the *Erebus* and the *Terror*, were waiting in Lancaster Sound, at the eastern entrance to the Parry Channel, for the pack ice to melt, Franklin met with the captains of two whaling ships, who were also waiting. Then the expedition's ships and the whalers went their separate ways.

Franklin's two ships entered the Parry Channel, and then disappeared without a trace. By 1847 those in England began discussing search-and-rescue operations. Sir John Richardson, knighted in 1846, who had served with Franklin on those two earlier expeditions, volunteered to lead the operation. But Richardson, too, was nearing 60 years of age. He had not traveled in Canada for over 20 years. No longer actively exploring, he was serving as an administrator at the Royal Hospital Haslar. He was one of the nation's leading naturalists but had not intended to return to the Arctic. He knew that in order to lead an expedition that would rely to a great extent on travel by foot he would need an experienced Arctic explorer to support his efforts.

Also in 1847, Rae returned to England on leave from the Company and published a brief report of his expeditions in *The Times*. When Richardson read it, he jumped to his feet and

---

* Now Tasmania

cried out to his wife, "I have found my companion, if I can get him!" Richardson had received hundreds of applications from men asking to be allowed to go with him as his assistant. However, he immediately contacted Rae to offer him the position of second-in-command of the proposed expedition.[279]

Rae spent very little time considering the offer, as he had numerous reasons to accept. First, from Franklin's earlier writings, he knew of Richardson, and had admired the stories of his exploits. He knew, too, that Richardson was considered one of England's most noted naturalists, who had published extensively following his earlier exploring voyages to the Arctic and Upper Canada. His summary of the proposed expedition would certainly be published upon his return to England. Finally, the search for Franklin was the talk of the entire country, and for Rae to be present when the riddle was solved would further advance his already growing reputation as an Arctic explorer. While Richardson secured permission from both the Admiralty and the Company for Rae to join the search, Rae wrote a very apologetic letter to Sir George Simpson, asking forgiveness for having accepted the assignment without the two of them having first discussed it.

In March 1848, Richardson and Rae left Liverpool on the steamer *Hibernia*, and after landing in New York they traveled on to Montreal by steamship. Sir George apparently harbored no ill will, as the two explorers were his invited guests for three days before departing for Sault Ste. Marie, where they met the 20 servicemen who would accompany them on their search. Over the next 14 months the group continued the search for Franklin, covering hundreds of miles on foot and by boat, and spending the winter living off the land. In May 1849 Richardson returned to England, leaving Rae and eight men from the original party to continue searching.

Franklin had now been missing for five years, and what had begun as a rescue mission had now evolved into a recovery effort to try to locate any vestige of their presence that might lead to determining what had become of the lost crews. Rae was frustrated, as the efforts to date had produced no results, and he was ready to return to England. However, while he was serving as the officer in charge of the Mackenzie River district for the Company, he received a dispatch from Sir George, stating "Her Majesty's Government had asked for the loan of your services" to continue the search for Franklin.[280] As a result Rae served as the head of the expedition, continuing the search, from the time of Richardson's departure until the spring of 1852. Although he was able to explore and chart hundreds of miles of new territory, the result of his efforts was the same: the *Erebus* and the *Terror*, and all the men who had been aboard those ships, seemed to have disappeared completely.

Rae continued his search for Franklin during the winter of 1851–1852. Then in March 1852 he sailed from New York, arriving in London in April. He made reports to both the Company offices in London and to the British Admiralty. Although years had passed since Franklin's disappearance, and Rae had only found two wooden spars which may or may not have been from one of the vessels, he found that interest in Franklin was greater than ever, and public opinion was certainly on his side. In May the Royal Geographical Society awarded Rae its Founder's Gold Medal:

for his survey … under the most severe privations in 1848 and for his recent explorations on foot and in boats … by which many important additions have been made to the geography of the Arctic regions." [281]

Rae enjoyed the time in his homeland and joined in the London social whirl for the entire summer. He also made several trips to northern Scotland, where he was a guest on the estates of wealthy landowners. He extended many of these visits by taking the opportunity to see his now widowed mother in Orkney. In the fall he took time for a visit to Paris, but he was getting restless. Toward the end of March 1853, he sailed for New York on the *Europa*. Soon after that he was back at Sault Ste. Marie, preparing to embark on the expedition that would finally solve two of the greatest mysteries of the day—locating the Northwest Passage and determining the fate of the Franklin Expedition.

Rae continued his methodical search for both the connecting link for the Northwest Passage and anything related to the Franklin Expedition. On March 31, 1854, after spending another Arctic winter hunting for his food, he and his party set out from Repulse Bay. Traveling by dogsled, Rae encountered an Inuit man wearing a gold cap braid as a headband. Rae learned that the man had traded for it, and that it had come from "the place where the dead white men were." Later that day Rae recorded the following in his field notes.

> Met a very communicative and apparently intelligent Eskimo; had never met whites before but said that a number of Kabloonas [white men], at least 35–40, had starved to death west of a large river a long distance off. Perhaps 10 or 12 days' journey. Could not tell the distance, never had been there, and could not accompany us so far. Dead bodies seen beyond two large rivers; did not know the place, could not or would not explain it on a chart.[282]

With no way of knowing which rivers the Inuit was describing, or even the direction in which to proceed, Rae returned to the Company's portion of his mission and continued his search for the Northwest Passage. He and his party forced their way north, following the route described by Sir James Clark Ross in 1830, eventually reaching a sea passage covered by what Rae described in his journal as "young ice." He realized that the surrounding land protected this channel from the thicker pack ice and that it would be navigable throughout the year, at times when the larger Victoria Strait was not. When Ross had charted the area, he had shown this as an enclosed bay; however, he had also drawn a dotted line and a question mark, indicating that it might include a navigable channel. Rae found that Ross's guess had been correct.

John Rae had finally discovered the missing connection—the only channel passable by the ships of that time—a passage between the Boothia Peninsula and King William Island. At that point the fact that this was the only possible passage was mere speculation, based on Rae's accumulation of knowledge and experience acquired over the past decade of exploration, but half a century later Roald Amundsen would travel the length of the channel and prove him correct. John Rae had discovered the final link in the Northwest Passage.[283]

Following his discovery, Rae and the party returned to Repulse Bay, arriving on May 26,

1854. They were met by a party of Inuit who had come to trade relics. Looking at what they had brought to trade, and having an opportunity to hear their stories, Rae realized they were describing the fate of the Franklin Expedition. The Inuit explained that four winters earlier some of their relatives had been hunting seals near the northern shore of what Rae determined must have been King William Island. The hunting party had encountered about 40 Kabloonas dragging a boat and some sleds southward. The next day the white men had turned east, toward what Rae recognized by its description as Back's River. Then the following spring, when the Inuit had returned to the area, they discovered tents and an overturned boat, each containing the bodies of dead white men. The Inuit also indicated, through their descriptions of the mutilated state of some of the dead, that the last survivors must have turned to cannibalism in a final effort to stay alive.

Rae spent days listening to their stories, and finally turned to the items they had come to trade. There were pieces of telescopes and guns, compasses, and watches. Some items bore crests or initials. By matching the markings against the ships' logs listing the names of officers from Franklin's two vessels, Rae was able to verify that the items were from the *Erebus* and the *Terror*. Several of the items could be linked to John Franklin himself; there was a gold watch and several eating utensils that bore the Franklin crest, and finally there was a silver plate engraved "Sir John Franklin, K.C.H." After years of searching, but within a period of less than three months since his Northwest Passage discovery, John Rae had now solved the second Arctic mystery and had obtained proof regarding the destiny of the Franklin Expedition.

Figure 8.4 John Rae
(Royal Geographical Society)

None of the Inuit with whom Rae had spoken, and from whom he had purchased the relics, had, they said, seen the "white men" either before or after they had died. In addition, none of these Inuit had visited the place where the bodies had been found, and all of their information and the artifacts had been relayed to them by others. Rae was not overly concerned by this, as the items they had produced could only be linked directly to the Franklin party. He also believed the Inuit had no reason to lie and believed their story in total. He was also certain that the crew members of both ships were all dead and had been for at least four years. He based this on the fact that he had "offered the Inuit guns, knives, saws, and kettles—everything they most valued, if they could tell of even one man, or the possibility of one man, being alive." [284] They could not, and again indicated that all of the white men were dead.

Rae felt he now had concrete proof of the fate of the lost expedition. He also had an important decision to make. His party had more than enough supplies to continue their exploration. But he also knew that at least three different groups were also looking for answers to Franklin's disappearance and, based on his newly discovered information, all were searching in the wrong places. **(Figure 8.4)** He needed to tell others quickly what he had found.

Although he had no doubt about any aspects of the Inuits' story, the evidence he had obtained could to a large extent be described as circumstantial. The items for which he had traded were clearly from the two ships—but as to the location of the bodies, whether any of the survivors could still be alive, and most importantly the indication of cannibalism, Rae was relying totally on the word of the Inuits. Upon his return to England this would become a contentious issue and would create dire consequences for this courageous Arctic explorer for the remainder of his life.

Great honors and powerful enemies

On August 4, 1854, Rae sailed out of Repulse Bay on the *Prince Rupert*, and back to England. [285] He had prepared two accounts, generally identical in their wording, one for Sir George at the Company, and the second for the Admiralty. Both reports included the stated suggestions of cannibalism. While crossing the English Channel he had also written a note to the editor of *The Times* describing the artifacts he had collected and indicating they were proof of the fate of the Franklin Expedition. In that letter, he was careful not to mention the assumption of cannibalism.

It is unclear how the information from what Rae had assumed would be confidential communications with the Admiralty and the Hudson's Bay Company found their way to that newspaper editor's desk. But the day after his arrival in London, and shortly following his initial meeting at the Admiralty, a lengthy article appeared on the front page of *The Times*. In addition to a description of the relics, it stated:

> From the mutilated state of many of the corpses and the contents of the kettles, it is evident that our wretched countrymen had been driven to the last resource—cannibalism—as a means of prolonging existence.[286]

The Admiralty welcomed Rae's report. It had spent over £600,000 on 55 separate search efforts pursuing leads regarding Franklin's disappearance. This report conclusively settled the matter, and it could begin putting this ongoing funding to better use. But from virtually every other quarter the response could not have been more negative. Rae's inadvertent revelations that the final survivors of the Franklin Expedition had resorted to cannibalism universally shocked Victorian England to its core. After the *Times* article had appeared, Rae paid a courtesy call on Lady Franklin, but she refused to even offer him a chair. She angrily challenged his information and said, "Such allegations should never have been committed to paper." Rae attempted to explain, but she ordered him to leave her home.[287] Soon the entire country joined in the backlash. Both *The Times* and *The Sun* rejected the thought of cannibalism—*The Times* because the "Inuit were all liars" and *The Sun* because no one of "this noble band of adventures would ever resort to such a practice."

Privately, Arctic veterans such as Sir John Richardson, Sir John Ross, and Leopold McClintock accepted the truth of Rae's report. Publicly, they all turned against him. Lady Franklin, vehement in her outrage, turned to a powerful ally and friend, the extremely well-known Charles Dickens. His popular weekly magazine *Household Words* provided him with the perfect venue to devastate John Rae and vindicate Franklin. In it, Dickens rabidly attacked Rae's report, its conclusions, and Rae personally. The explorer issued a response, but his writing skill was no match for that of the famous author. To make matters worse, many of the remarks in his rebuttal infuriated the Royal Navy, which up to that time had been supportive. Rae's steadfast defense of the Inuit and the overall accuracy of his report, over the objections of the powerful Lady Franklin and her confederate Charles Dickens, further alienated an already biased English society.

Not only was Rae's personal reputation in tatters, but he began to suffer financially as well. A reward of £10,000 for proof of the fate of the Franklin Expedition had been offered, but he had not known of it until notified by the Admiralty. But when, at the Admiralty's request, Rae applied for it, Lady Franklin increased her attacks against him. The Admiralty, yielding to her, announced that the money would be held until Rae's story could be verified by an overland search expedition. Meanwhile, the Company, also giving into the public sentiment, declared that Rae's regular pay would be suspended until the Admiralty agreed to pay the reward. Ultimately some form of justice prevailed, and both the Admiralty and the Company paid Rae the money owed. But Rae neither forgot nor ever forgave either organization for the ordeal to which he had been submitted.

In 1859 an overland expedition was conducted by the Admiralty, and while searching in the area described by the Inuit, they discovered a cairn.* It contained the only surviving written report created by the missing expedition's members. This document, along with subsequent discoveries made years later, completed the story. Early in the expedition, both the *Erebus* and the *Terror* had become icebound and, the men being unable to free the vessels, the pressure from the expanding ice had destroyed both ships. The crew members were forced to abandon them and attempt to reach safety by traveling overland, but quickly began to succumb to the elements. Franklin had been among the first to die, and within a year his death was followed by that of 9 other officers and 15 crewmen. The survivors had then headed south toward Back's River. Tragically, the entire party of about 130 men had died from a variety of causes: some of hypothermia, some from scurvy, and many from starvation during their attempted overland trek southward. Everything discovered by the Admiralty's expedition served to confirm the information provided by the Inuit. Rae's account had been proven completely true. But this verification would not be released until years after his death.

---

* A mound of rough stones built as a memorial or landmark

Canadian return

In 1856 Rae resigned from the Company after 23 years of employment. However, his passion for travel and adventure was far from extinguished. With some of the funds ultimately provided by the Admiralty he decided to prove his theory regarding the location of the Northwest Passage by sailing through it from the Atlantic Ocean to the Pacific. Moving to Hamilton,* and securing the assistance of two of his brothers, Rae began the construction of a schooner that he felt would be up to the task. The ship was completed, but too late in the season to attempt the journey. The three brothers, agreeing that they would not let their investment sit idle, began using the vessel to haul cargo on the Great Lakes. But then on a voyage from Cleveland to Kingston, Ontario, the ship was lost in a storm, and Rae's plan to travel the length of the Northwest Passage was lost with the ship.

Rae continued to live in Hamilton until 1859. He traveled extensively throughout Canada and the United States, often being asked to give presentations on Arctic exploration to various groups. During one of his speaking engagements he met Catherine (Kate) Thomson. Her father, a major in the British Army, did not initially welcome this prospective son-in-law. Among other issues, Rae was 46 and Kate was 21. But after his initial resistance Major Thomson gave his blessing, and in January 1860 John and Kate were married in Toronto. They left for England on their honeymoon, planning to return to Canada. However, they would spend the rest of their married lives in England and Orkney and would never again live in Canada.

Rae continued to pursue his exploring and surveying interests. The Atlantic Telegraph Company wanted to create a telegraph line that would link America and Britain. This involved installing a line from the Scottish mainland to Orkney, to Shetland, to the Faroe Islands, to Greenland, to Iceland, and on to Labrador. Rae was contracted to conduct the land portion of the survey, and he completed the survey work for the Faroe Islands, Greenland, and Iceland before returning to London in 1861. Again, he was employed by the Company, this time in the construction of a telegraph line across the entire reach of Canada, and this he accomplished by traveling the full distance once more, much of it on foot. Then in 1882 he visited Canada for the last time; he went as a guest of the American Association for the Advancement of Science, where he presented a paper on Arctic Exploration in North America.

Going full circle

In the 1880s Rae began having health problems. This did not stop him from rising early each morning to begin his day with an extended hike. He also continued to directly confront his critics head on, spending considerable time and great effort in contacting writers and publishers who he felt had used incorrect information, or credited the wrong individuals, in their writings. In 1893 he began suffering from lung congestion and on July 22, at age

---

* City in Ontario, Canada between Lake Ontario and Lake Erie

79, John Rae passed away. He had left no instructions on a final resting place. But Kate, his wife of 33 years, felt he should return to Orkney. He is buried in the churchyard behind St. Magnus Cathedral in Kirkwall, the capital of the Orkney Islands. Inside the cathedral is a larger-than-life Portland stone sculpture of the explorer sprawled under a buffalo hide, wearing Inuit mukluks* and with an open book and musket by his side. (**Figure 8.5**) The inscription reads:

John Rae, M.D., L.L.D., F.R.S., F.R.G.S.

Arctic Explorer

Intrepid discoverer of the fate of Sir John Franklin's last expedition

Born 1813—died 1893

Expeditions: 1846–7. 1848–9, 1851–2, 1853–4

Erected by public subscription, 1895 [288]

To sum up, Rae, the son of a middle-class land manager, became an agent for the Hudson's Bay Company. He had been trained as a physician, but like many others was called to a life of adventure. He was a man of courage and principles. His physical exploits show him to have been a man of almost superhuman physical stamina. He solved the two great mysteries of Victorian Arctic exploration by discovering the final link in the Northwest Passage and ascertaining the outcome of the Franklin Expedition.

In 1903, Roald Amundsen led the first expedition to traverse Canada's Northwest Passage. His work confirmed that John Rae had successfully discovered the channel that linked the Atlantic and Pacific oceans.

In the 1980s, many years after Rae's death, autopsies conducted on the remains of the missing crew members from the Franklin Expedition were analyzed. Using methods not available at the time of the men's disappearance, scientists determined that the ships' crew had succumbed to starvation as well as suffering from scurvy and tuberculosis. Additional tests showed that the tins of preserved meat that had been taken onboard in Greenland had not been sealed properly, so lead from the sealant had leached into the contents and many of the crew members may have died from lead poisoning.[289] The wreck of the HMS *Erebus* was finally discovered in 2014, and the HMS *Terror* was found in 2016. Between 2021 and 2024 researchers extracted DNA materials from the bones of the fallen officers and crew members. To date the remains of only two of the expedition's members have been linked to living descendants.[290] As for the dispute over cannibalism, forensic evidence, including the cut marks on the skeletal remains evaluated by contemporary researchers, has vindicated Rae and his Inuit observers in every detail.[291]

In the end, the social position and political standing of Lady Franklin would appear to have won out. Rae's steadfast loyalty to the Inuit and his decision to accept their reports as accurate cost him dearly; through her very public onslaught his reputation was publicly challenged,

---

* Soft, thickly insulated, boots, made of caribou or seal skin, worn by the Inuit

his finances were threatened, and all attempts to honor his many accomplishments were systematically beaten down. Ken McGoogan says:

> When prevarication might have kept a coveted knighthood within reach, Rae stood by his principles, refusing to recant, refusing to sell out the Inuit people. And so this peerless figure became the only major British explorer of the age never to receive a knighthood.[292]

Against intense pressure from within the highest echelons of British society, Rae refused to turn against his personal beliefs or against the Inuit, whom he considered friends. Although decades would pass, in the end his discoveries would be proven conclusively. Perhaps it is John Rae who actually carried the day.

Figure 8.5 John Rae Memorial (Orkney Library Archives)

9

Sir Charles Wyville Thomson (1830–1882)

My soul is full of longing for the secret of the sea,
and the heart of the great ocean sends a thrilling pulse through me.

–Henry Wadsworth Longfellow

As the British Imperial Century progressed, so too did the nature of exploring. The earliest efforts had focused on natural science, in the belief that a finite number of plants and animals, known as "constant typological classes," existed—and, importantly, had existed since the Creation. It was believed that if expeditions could travel to the far corners of the world, then eventually all the varieties of flora and fauna could be collected, studied, classified, and catalogued. The explorers who had been sent forth under the direction of individuals such as Joseph Banks had collected plants and developed the concept of economic botany upon which much of the British Empire had been built.

By the last third of the 19th century the fact that the empire had been largely explored changed the nature of exploring— but it was not the only factor. What was crucial to the change was the break from the concept of constant typological classes that emerged as a result of Charles Darwin's theory of evolution through natural selection. **(Figure 9.1)** His theory[*] was presented in his groundbreaking *On the Origin of Species*, published in 1859. The work upended the study of natural science.[293] Faced with the concept of an ever-changing number and variety of plants and animals, and the prospect that what had been believed to be nature's clearly defined boundaries might no longer exist, interest in natural science as a separate scientific discipline, as well as a focus of the general public, quickly waned. Voyages such as those undertaken by Menzies rapidly became a relic of the past.

Exploration nevertheless continued, now concentrating on increased trade and better utilization of the territorial acquisitions within the empire. Commercial explorers also continued to create maps and charts that allowed traders to enhance Britain's already strong mercantile position. Although Darwin's voyage aboard HMS *Beagle* from 1831 to1836 and

---

[*] Which he called "descent with modification" rather than "evolution"

Figure 9.1 Charles Darwin (Wikimedia Commons/PD)

Huxley's on HMS *Rattlesnake* from 1846 to 1850 resulted in significant scientific breakthroughs, the primary objectives of both expeditions had been commercial exploration to enhance trade and territorial annexation.[294] The two naturalists' revolutionary discoveries were simply a byproduct of those primary goals. In addition, private ventures aimed at the suppression of slavery and the spread of Christianity had now moved from exploring missions into the realm of implementation through governmental bodies and missionary associations. Kirk's work in the eradication of slavery in East Africa certainly had a humanitarian basis but was also designed to replace the trade in human cargo through the creation of a market where British manufactured goods could be exchanged for African gold, ivory, and palm oil. While Rae had established the fate of the lost Franklin Expedition, the object for both Franklin and Rae traveling in the Arctic was to discover the Northwest Passage, which would enhance the commercial link between Great Britain and the Far East.

Although commercialism is an understandable goal for an empire, it was not enough to support ongoing efforts in exploration. In the final days of the British Imperial Century, with most of the earlier reasons diminishing, the emphasis changed again. The goal of the medical explorers moved from exploration in support of commercial advantage to voyages simply for the sake of knowledge, and in the process they would create an entirely new branch of science.

Down to the sea in ships

They that go down to the sea in ships, that do business in great waters. These see the works of the LORD, and his wonders in the deep. For he commandeth, and raiseth the stormy wind, which lifteth up the waves thereof. They mount up to the heaven, they go down again to the depths; their soul is melted because of trouble. They reel to and fro, and stagger like a drunken man, and are at their wits' end. They then cry unto the LORD in their trouble, and he bringeth them out of their distresses. He maketh the storm a calm, so that the waves thereof are still. Then they are glad because they be quiet, so he bringeth them unto their desired haven." Psalm 107: 23–30 KJV.

About 71 percent of the Earth's surface is covered with water, and humanity has been both attracted to and fearful of the seas and oceans since prehistoric times. Until comparatively recently, successful voyages or those that failed—that is, a safe return or being lost at sea—were seen as acts of God: unfathomable. Initially, knowledge about the great bodies of water was limited to observations of their surfaces and the study of the animals that fishermen retrieved in their nets. Today, this would be called marine biology, the study of marine organisms: their physiology, characteristics, and life history. However, it was not until the writings of Aristotle and Strabo (author of *Geographica*, which presented a descriptive history of regions of the known world) between 384 and 322 BCE, that specific references to marine life were recorded. Aristotle identified a variety of species, and developed the initial classifications for the various crustaceans, echinoderms, mollusks, and fish that he collected. Because he was the first to record observations on marine life, he is often referred to as the father of marine biology.[295] Modern science is often said to stand on the shoulders of Aristotle, and in this case, it is most certainly true.

While the study of the organisms that lived within these great bodies of water remained at a preliminary stage, the study of the ocean's surface was moving forward at a rapid pace. As early as 1200 BCE the Phoenicians began utilizing celestial navigation for their ocean voyages. When the Portuguese began navigating the Atlantic Ocean in their search for a route to the spice treasures of India, they established the earliest example of a systematic scientific project to study the ocean. Their early exploration initially involved only cartography, but it was soon expanded to include the study of currents and winds. The charts, maps, and navigational methods developed by the Portuguese were much more advanced and more commercially viable than those of their exploring rivals. These documents were so prized that they were held in the tightly guarded Royal Archives in Lisbon and carried the death penalty for anyone discovered disclosing the information.[296] The methods developed by the Portuguese were eventually replicated by the Spanish, Dutch, and English.

But well into the 19th century knowledge of the oceans remained largely confined to navigational needs and to marine life that could be found in the uppermost fathoms of the water. It was only through the efforts of a small number of men that the creation of maps to navigate the surface evolved into a science that would conduct investigations from the tops of the waves to the very bottom of the deepest seas, and their leader would spend his brief lifetime creating the parameters for what this discipline was to become.

Medicine to natural science to paleontology

Charles Wyville Thomson (**Figure 9.2**) was born at Bonsyde, in Linlithgow, West Lothian, on March 5, 1830. No comprehensive biography of Thomson has ever been published, but from the meager accounts of his life that do exist, a brief narrative can emerge. His father, Andrew, had been a surgeon in the service of the East India Company, and his mother, Sarah

Figure 9.2 Sir Charles Wyville Thomson (Wikimedia Commons/ PD)

Ann Drummond Smith, had moved from India to Scotland for the birth of their son. This separation of his parents continued through much of Thomson's youth; however, even from a distance he felt his father's influence, in that his academic expectations were high, and there was an anticipation that Charles would excel in his studies. There is no indication that this pressure to succeed had any negative effects, and by all accounts the boy was quite happy at school and excelled in his studies. But as the pre-university curriculum at that time concentrated on the classics and there was little formal instruction in the sciences, he supplemented this learning through his own efforts, roaming the countryside on weekends and holidays. He evolved into an inveterate collector and observer of nature, and natural sciences became his lifelong passion.[297]

When Thomson entered the University of Edinburgh, perhaps at the urging of his father, he enrolled in the medical school. He completed his studies in medicine, but early in his university education decided to focus entirely on the natural sciences as opposed to a career as a doctor. In 1847, while still a university student, he was accepted into the Botanical Society of Edinburgh, and he began attending the botany classes conducted by John Hutton Balfour. In 1850 he left the university without taking the examinations required for a medical degree, yet already with an established reputation in natural history, and through contacts established within the Botanical Society, his academic ascent was both impressive and rapid.

In 1850 Thomson was appointed Lecturer of Botany, and in 1851 Professor of Botany at the University of Aberdeen. In 1853 he became Professor of Natural History at Queen's College, Belfast, and a year later he was appointed to the Chair of Mineralogy and Geology at Queen's University, also in Belfast. He had always been interested in the ocean and had spent many hours exploring the Scottish and Irish coastlines, but up to the time of his appointment as chair of mineralogy and geology his curiosity had been related to the biology of living things. While at Belfast he taught courses in botany and zoology, which were directly linked to his background in natural science. But he also was asked to teach geology, and this evolved into what would become his lifelong interest in paleontology.

Thomson began to evolve from a naturalist studying land-based living plants and animals into a marine biologist specializing in studying fossils that had come from the sea. He published studies of living creatures, including coelenterates and polyzoans, and fossilized remains of cirripeds, trilobites, and crinoids.* Recognition of his work resulted in his being elected a Fellow of the Royal Society of Edinburgh in 1855. This change in direction also prompted him to pose a question that would set the focus of his career for the remainder of his brief life: "Could these ancient organisms, found on land only as fossils, be

---

* Coelenterates are jellyfish/corals/sea anemones; polyzoans are small, mosslike creatures; cirripeds were barnacle-type creatures; trilobites were marine arthropods, and crinoids (some of which still exist) are sea lilies and are related to starfish.

found alive in the depths of the oceans?" A second question, "To what depth in the ocean can life exist?" would place him at odds with his alma mater and with one of the most distinguished naturalists of his time.

The *Lightning* and the *Porcupine*

Edward Forbes was born on the Isle of Man, and like Thomson studied at the University of Edinburgh. Also, like Thomson, Forbes moved from the study of medicine into the discipline of natural science. He participated in a number of exploration voyages, compiled a distinguished publication record, and held a number of scientific and academic appointments before ultimately being named Professor of Natural History at the University of Edinburgh. During his career, Forbes proposed many scientific theories, and he is remembered for two of these hypotheses in particular. He had proposed that distributions of plants and animals of the same species found on now-isolated islands were a result of separation of sections of land from the mainland during the Ice Age.* Now surrounded by water, the animals and plants formerly found on the mainland continued to exist on the islands that had been formed eons earlier. His theory was that the animals had not relocated, but instead that the land upon which they had always lived had moved. This mechanism, the first to explain the remote distribution of the same species, was discovered independently by Charles Darwin, who then credited Forbes with the findings. But not all of Forbes's theories would be validated.

During the few months that Forbes held the Chair in Natural History at Edinburgh, he also formulated what became known as one of the strangest scientific ideas to emerge from Victorian England. He called this the "azoic zone theory," † and it postulated that below 300 fathoms‡ no life could exist in the oceans. The professorship in natural history at Edinburgh University was certainly the most influential post in the natural sciences in Great Britain, and perhaps unsurpassed any similar post anywhere in the world at that time, so pronouncements made by the holder of this eminent position were treated as absolute truth. Edward Forbes died in 1854, within months of his pronouncement of the azoic zone theory, and 15 years later this theory was still a commonly held belief within the scientific community.

But Charles Wyville Thomson felt certain that Forbes's azoic zone theory could not be correct.[298] In 1866, while visiting the Norwegian biologist Michael Sars, Thomson examined the evidence of organic life dredged from the bottom of a fjord over 430 fathoms§ deep. Later, in discussions with Fleeming Jenkin, the University of Edinburgh's first Professor of Engineering, Thomson had been told that frayed ends of a telegraph cable retrieved from the bottom of the Mediterranean Sea, well below 300 fathoms, were covered in barnacles. Did this mean that the azoic zone theory was correct, but not at the depth Forbes had estimated? Or did it mean that the entire theory was without merit, and that life existed even in the ocean's greatest depths? As Richard Corfield states (2003), Thomson had "prestigious and influential

---

* At the time it wasn't apparent that more than one ice age had taken place.
† 'a' = without, and 'zoic' = life
‡ 1,800 feet/550 meters
§ 2,600 feet/800 meters

friends," and to secure an answer he turned to a friend in one of the highest positions in Victorian science.[299]

Thomson was determined to investigate Forbes' theory systematically, and to accomplish this he turned to William Benjamin Carpenter, a friend who shared an interest in paleontology and with whom he had worked earlier in his career. Carpenter was Vice-President of the Royal Society. As a senior officer within the most prestigious scientific body in the world, Carpenter held an unusual position; he was a bureaucrat with access to funding and equipment not available to a simple academic — but having been elected to the Society based on his accomplishments in science, he also had the credibility to ensure that his proposals would be received with respect. Carpenter was persuaded by his friend Thomson that the azoic zone theory was worth investigating. He was also persuaded to approach the Admiralty to secure Thomson's use of the steam frigate HMS *Lightning* for the summer of 1868.

Throughout the summer the *Lightning* sailed between the Shetland and Faeroe Islands. Carpenter and Thomson were joined by John Gwyn Jeffreys, a Welshman who was also a member of the Royal Society and an expert on conchology.* Jeffreys had developed a successful technique for dredging the seabed on his earlier expeditions, and he now introduced this procedure to his shipmates onboard the *Lightning*. Although the ship had not been designed for scientific research, during that summer the trio made two extremely significant discoveries. First, their dredging operations found that animal life in the form of marine invertebrate groups existed at a depth of more than 600 fathoms.† Life at that depth had never been considered; it was twice the depth at which Forbes had predicted the azoic zone would begin. Second, when compiling readings of physical characteristics at various depths, the men discovered that the deep ocean below 200 fathoms‡ was dominated by currents that produced oceanic circulation and temperatures substantially different from those of the surrounding water.

This second discovery was even more significant. It was known that the oceans are in constant motion. This could be observed by watching the waves crash onto the shore or by experiencing surface currents, such as the Gulf Stream, moving water across the ocean just as rivers move water across the land. It was known, too, that surface currents are powered by the various wind patterns and that they fluctuate based on conditions found above the surface of the water. But Thomson had discovered that there were also deep underwater currents. The process became known as thermohaline circulation. Although the ocean's deep currents move at a slower rate than the surface currents, thermohaline circulation moves massive bodies of water around the globe, from northern oceans to southern and back again. In addition, there are currents that slowly turn over water in the entire ocean, from top to bottom; in certain areas they move warm surface waters downward, and in others they force cold, nutrient-rich waters upward, mixing the oceans' waters on a global scale. Thomson also determined that these currents operated independently of the conditions on the surface, unaffected by them.

The voyage of the *Lightning* was so successful that it was followed by expeditions utilizing

---

* The study of mollusk shells
† 3,600 feet/1,100 meters
‡ 1,200 feet/ 366 meters

the HMS *Porcupine* in the summers of 1869 and 1870. These expeditions investigated the deep seabed to the west of Ireland. The dredging operation found a wide range of sea life existing at a depth of over 1,600 fathoms,* far deeper than their earlier work. Thomson describes this discovery in *The Depths of the Sea* (1873) which chronicles the voyages of both the *Lightning* and the *Porcupine*.

> For the bed of the deep sea, the 140,000,000 square miles which we have now added to the legitimate field of Natural History research, is not a barren waste. It is inhabited by a Fauna more rich and varied on account of the enormous area, and with the organisms in many cases apparently even more elaborately and delicately formed, and more exquisitely beautiful in their soft shades of colouring, and in the rainbow-tints of their wonderful phosphorescence, than the fauna of the well-known belt of shallow water.[300]

This conclusively proved that as marine life existed in the deepest of ocean waters, the azoic zone did not occur at any depth. The scientists on the *Porcupine* also made an extensive study of the North Atlantic Current of the Gulf Stream. They verified that it was responsible for the temperature variations of the seas along the western coast of Ireland and along the western coastal islands of Scotland, which are consistently warmer than the seas of the eastern coasts. For his efforts, Thomson was made a Fellow of the Royal Society in 1869. Then in 1870, having not yet reached the age of 40, he was appointed Professor of Natural History at the University of Edinburgh, the position Edward Forbes had held in 1854.

Preparing for the Challenger Expedition

With the success of the Lightning and Porcupine expeditions, combined with Thomson's appointment to the Chair of Natural History at the University of Edinburgh and Thomson's and Carpenter's positions within the Royal Society, the two men now had the necessary influence to persuade the Council of the Royal Society and the Admiralty to organize a much larger expedition. At the request of Thomson and Carpenter, the Royal Society agreed to sponsor the expedition and formed a Circumnavigation Committee. This group of prominent scientists developed the projected course the ship would follow, and also proposed a fourfold mission for the proposed expedition:

1. To investigate the physical conditions of the deep sea in the great ocean basins (as far as the neighbourhood of the Great Southern Ice Barrier) in regard to depth, temperature, circulation, specific gravity, and the penetration of light.

---

* 9,600 feet/2925 meters

2. To determine the chemical composition of seawater at various depths from the surface to the bottom, the organic matter in solution, and the particles in suspension.

3. To ascertain the physical and chemical character of deep-sea deposits and the sources of these deposits.

4. To investigate the distribution of organic life at different depths and on the deep seafloor.[301]

The Admiralty, not noted for its speed in approving special projects, worked with surprising momentum and generosity. Gladstone's government agreed to subsidize the voyage, and the application for funding created by the Royal Society was approved in late 1871. The Admiralty committed HMS *Challenger*, (**Figure 9.3**)—a 70-meter corvette, the smallest class of vessel in the Royal Navy that can be classified as a warship—to the project. They agreed to cover the costs associated with converting a ship designed for traditional service within the Navy to one that would focus on a purely scientific expedition. They would also second approximately 240 sailors and officers from the Navy to the project, and they gave their final approval in April 1872.[302] By that December HMS *Challenger* was ready to sail. The ambitious plans would mark the Challenger Expedition as the costliest exploring venture ever undertaken, and it would be the first global oceanographic expedition driven purely by science. It is not surprising, then, that there were at least four additional reasons why the British Government was so keen to provide the equipment and manpower for this vast undertaking.

First, in the latter decades of the 19th century the economic supremacy of Britain was

Figure 9.3 HMS Challenger *(Wikimedia Commons/PD)*

unrivalled. Britain controlled the largest empire in the world, and it could afford to sponsor an expedition solely for scientific discoveries. In doing so it increased the empire's status, and in a very visible way enhanced its prestige. Second, the empire's wealth and superiority had been built on decades of maritime dominance. The Royal Navy had no difficulty in committing the resources necessary to make this venture happen when few other countries could do the same. The expedition was a very public show of this wealth and resources. Third, knowledge gathered through the study of the world's oceans would serve to maintain Britain's dominance over its maritime rivals. Finally, Darwin's theory of natural selection, published only 14 years earlier, was still being hotly debated across the empire. One of the most significant problems for Darwin was establishing a burden of proof. He had theorized that certain marine organisms which were found on land only as fossils could be found still alive in the oceans. If the investigation of the ocean's depths produced living organisms not found alive on land, it would validate both the concept of mutability in the plant and animal worlds and the traits of constant change and nature's transitory quality. So, the Challenger Expedition would not only investigate the geology and geography of the sea and seafloor, but also attempt to prove the theory of "descent with modification," and in the process conclusively validate Darwin.[303]

HMS *Challenger* was launched on February 13, 1858. The ship could be termed a "transition vessel," spanning the shift between steam and sail. It was a three-masted vessel fitted with a steam-driven propeller assembly that could be raised when the ship was traveling under sail. It was still essentially a sailing ship, though, and it would travel under sail for most of the voyage. The steam engine was used mainly to power the dredge used for collecting samples. At 230 feet in length and 40 feet across the vessel was not a large one, especially for the substantial crew and the abundance of equipment necessary for the extensive voyage. Its commanding officer was Captain George Nares, a Scotsman who in later years was knighted for his work in Arctic exploration, and under him were about 20 naval officers, including surgeons and engineers. Between its launching and its reassignment for Thomson's project the *Challenger* had traveled to America and the West Indies, Australia, New Zealand, and Fiji on various military assignments. Now, in preparation for its upcoming assignment, all but two of its 17 guns were removed to make room for the laboratories, storage cabinets, extra cabins, and a special dredging platform that would be needed for its extensive scientific mission.[304]

The science of oceanography is born

On December 21, 1872, six civilian scientists and the large Royal Navy contingent departed Portsmouth for a journey that would occupy the next three and one-half years. The expedition they embarked on was, uniquely, a product of its place and time. Large blocks of time would be spent in the various ports of call, but the ship would spend 713 days of the voyage at sea, and travel 68,890 nautical miles.[*]

---

[*] The land mile, or statute mile, is 5,280 feet, and is based on paces, whereas the nautical mile, used for distances on the ocean, is a mathematical calculation based on degrees of latitude around the equator, each degree measuring 6,070 feet. So, in its three and one-half years the *Challenger* traveled nearly 80,000 statute miles.

Each of the six scientists had been selected for the various areas of expertise that he could bring to the mission. Soundings and temperature measurements would be taken, and samples of the bottom sediment, water, and biological specimens were all to be collected. These individuals shared several characteristics. They were all from families that could be considered at least moderately well off, and they had all pursued scientific studies at the university level but had developed their various scientific theories through their personal experiences rather than in the classroom. As Doug MacDougall states, "one of the characteristics of many highly curious people is a penchant for following their own interests rather than those imposed on them." [305] That is not to say that they had neglected their formal education, but they had clearly valued their own scientific endeavors over what they had studied in the lecture halls.

Not included among the scientists who would be participating in the voyage was William Benjamin Carpenter. Although he had been instrumental in promoting and participating in Thomson's previous voyages, and he was deeply involved in the preparation for the three-year circumnavigation, he would not be joining the Challenger Expedition. There was, and continues to be, speculation regarding his absence, and there are at least three possible theories that have been put forward. First, advancing years could have prevented the committee from selecting him. He was nearly 60 years of age, and a voyage of nearly four years was going to prove strenuous for even the younger members of the expedition. Second, at the time of the voyage he was serving as registrar for London University, and the committee may have felt that his participation in short-term voyages during the summer months was considerably different than his absence from his important administrative duties for nearly four years. Finally, he was clearly the senior member of the team. He was older than Thomson, had been appointed to the Royal Society well ahead of his colleague, and held the second most important position within the sponsoring organization. It could be that Carpenter felt that he would participate in the expedition either as its leader or not at all.

Perhaps it was a mixture of all three theories, but whatever the reason or combination of reasons, it was Thomson who had been selected as chief scientist for the expedition. Carpenter's expressions of anger and humiliation at not being selected made Thomson's last few days in Portsmouth extremely difficult. As he was preparing to leave, he wondered, in a letter to his wife, whether he had been right in accepting the committee's invitation to head the expedition in the face of Carpenter's wish to do so. It was a sad ending to a close friendship that had developed through their common interests and their collaborative work —which was permanently destroyed with the voyage of the *Challenger*.

When the expedition began, almost nothing was known about the nature of the deep ocean. Scientists' two primary methods for discovering what was beneath them were sounding and dredging. Sounding was used to measure the depth of the ocean through which a ship was passing, and for this purpose the *Challenger*'s gear included over 181 miles (291 km) of Italian hemp. Crew members would lower a hemp line with a weight, or sinker, attached to it until it reached the seafloor. The line was marked with flags in 25-fathom* intervals to denote depth. The sinker often had a small container attached to it that would allow for the collection

---

* 150 feet or 46 meters

of bottom sediment samples, and sometimes attachments placed on the sounding lines would measure temperature and the speed of the current at various levels.

The crew also used a variety of dredges and trawls to collect biological samples. The dredge was essentially a large mesh bag that was dragged along the seafloor and scooped up whatever it encountered on the bottom. Mop heads, attached to a wooden plank that moved ahead of the dredge, swept across the seafloor and dislodged organisms that were then caught in the bag. Dredging was a long and tedious process. In the deeper parts of the ocean, it could take an entire day to lower the dredge and drag it along the ocean floor, then slowly and carefully bring it back up again. Trawls were larger metal nets that were towed behind the ship to collect organisms at different depths.

Upon the retrieval of a dredge or trawl, the crew members would sort, rinse, and store the specimens, which were then catalogued by the team's scientists. Removal of the majority of the *Challenger*'s cannons had provided space for extensive laboratory facilities **(Figure 9.4)** below deck. Today these methods seem primitive, but at that time they represented state-of-the-art Victorian remote-sensing technology. Their findings would prove invaluable to the study of marine biology and would lead to the establishment of a new branch of science: oceanography.[306]

Figure 9.4 Challenger Lab (Wikimedia Commons/PD)

The ship sailed from Sheerness in December 1872, and during the course of the voyage the *Challenger* would round both the Cape of Good Hope and Cape Horn, and sail toward Antarctica as far as the Great Ice Barrier. En route, the crew established 360 stations where they measured the bottom depth, took the sea's temperature at various depths, made observations of weather and surface conditions, and collected samples of the seafloor, the water, and organisms. Stops were made in Nova Scotia, the Caribbean, and South Africa, before pushing on to the Pacific. The *Challenger* reached Hong Kong in December 1874. At that point Captain Nares and several of the crew and officers left the ship to take part in the British Arctic Expedition, and Captain Frank Tourle Thomson (no relation) assumed command. Their route then took them to Indonesia and numerous remote islands in the Atlantic, Pacific, and Southern Oceans. On March 23, 1875, the *Challenger* was at Sample Station 225, located between the islands of Guam and Palau, and there the crew recorded a sounding of 4,475 fathoms.* This chance location proved to be the southern end of the Mariana Trench, still one of the deepest known places on the ocean floor. Thomson and the other scientists must have been elated at locating such depths beneath the ocean's surface. From the western Pacific, *Challenger* headed north to Hawaii and south again before passing back into the Atlantic through the Tierra del Fuego archipelago. The homeward stretch took it through the Atlantic, into the English Channel and finally home, arriving in May 1876.[307]

---

* 26,850 feet or 8,184 meters

The premature end

The Challenger Expedition was deemed a great success, but the expedition also encountered its share of tragedy. The youngest member of the civilian scientists was Rudolf von Willemoes-Suhm, a German national; upon completing his PhD he had been appointed Lecturer in Biology at the University of Munich. He had been visiting Edinburgh at the time the participants were being selected, and he requested an interview. Having just turned 25 years of age, his was the final appointment to the six-man scientific crew. But in the third year of the expedition, as the ship was starting its homeward journey in the Pacific, the young scientist contracted a bacterial infection. Antibiotics could have quickly cured him; however, their use was still well into the future. In spite of the best efforts of the ship's surgeons, he died and was buried at sea. Losses of a different type severely impacted the Royal Navy personnel assigned to the voyage. The ship's quarters were cramped, and many of the sailors were undoubtedly unimpressed by the importance of the voyage, so the constant labor associated with the dredging in addition to their normal duties would have provided a great temptation to desert. In the end, only 144 of the crewmembers remained of the nearly 240 who had departed from England; 7 had died and 26 had been invalided out of service or left at one of the hospital ports along the way. When Captain Nares had been reassigned at Hong Kong 5 expedition members had departed with him. The remaining absences were due to desertion, mostly to the newly opened gold mines in Australia.[308]

But there was also glory. On his return home Charles Wyville Thomson received a number of academic honors as well as being knighted by Queen Victoria for his service to science. The scientists from the expedition, who had also gained great prestige through their participation, began writing and publishing their accounts almost as soon as they returned to England. It was decided that the final official report of the *Challenger's* voyage would require multiple volumes, with much of the writing falling to the remaining five scientists who who, along with Thomson, had completed the expedition. In addition, however, specimens brought back by *Challenger* were distributed to the world's foremost experts for examination and written analysis, and while that process greatly enhanced the credibility of the final report, it significantly increased the expense and time required to finalize it. Activities were coordinated from the Challenger Office in Edinburgh, where Thomson served as head. The location allowed him to continue his teaching and administrative duties at the university.

He swiftly published a preliminary account of the initial portion of the journey, titled *The Voyage of the Challenger in the Atlantic*. Then he spent the next two years working on Volumes One and Two of the full narrative of the voyage. He was described as highly strung, and his health was generally poor throughout his life, so he found it enormously stressful to accord with the publishers' requirements to create numerous volumes of detailed illustrations and scientific descriptions. (**Figure 9.5**) The physical strain from the nearly four-year voyage, and the extensive and complex amount of work involved in coordinating the Challenger Office, coupled with his academic and administrative responsibilities at the university, did not allow him time to recover either physically or emotionally from his journey. While he had the heart of an explorer, sadly he had neither the physical nor emotional fortitude, and it soon became obvious to his colleagues that he was overextended. He tried to reduce his workload; in 1879 he ceased his university duties and in 1881 he gave up overseeing the reports of the expedition.

But it was too little, too late. On March 10, 1882, Charles Wyville Thomson took to his bed and died, a broken man, at Bonsyde. He had just turned 52 years of age.

John Murray, one of the six scientists who had been present on the *Challenger*, assumed responsibility for the official report. The findings from the expedition continued to be published until 1895, no less than 19 years after the completion of the voyage. When completed, the report, with all its necessary appendices, took up 50 volumes and over 29,500 pages.

To sum up, Charles Wyville Thomson was the son of a surgeon employed by the East India Company. Like the other medical explorers, he elected not to enter medical practice but to spend his brief life as an explorer. But unlike most of his contemporaries he did not travel as a naturalist, or to end slavery, or to advance

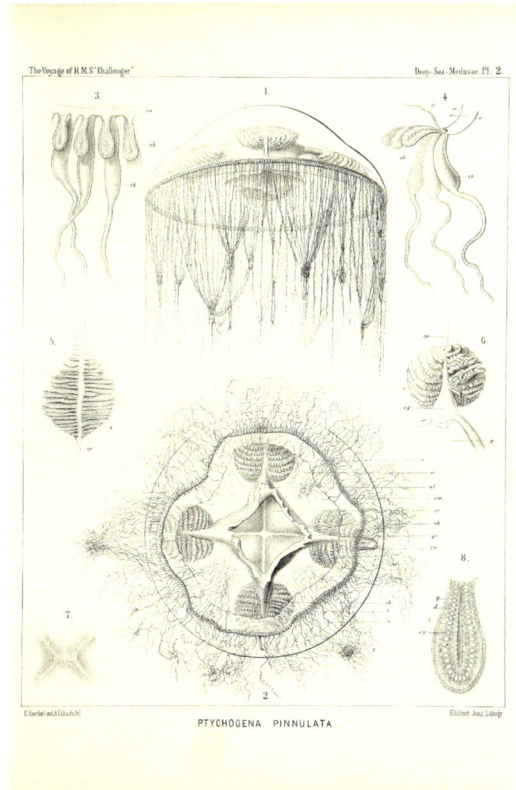

Figure 9.5 Challenger Scientific Illustrations(Wikimedia Commons/PD)

commercial interests. Instead, he devoted his efforts to the advancement of science. Before 1872, when the *Challenger* set out on its historic journey, the depths of the oceans were very much unknown. Marine science was mainly speculative, and the word "oceanography" did not appear in any dictionary. By trawling and sounding the North and South Atlantic and the Pacific oceans, traveling to the northern limits of drift ice in the Polar seas, and venturing far south of the Antarctic Circle, Thomson and his team discovered 4,700 new species of marine plants and animals, and completed a host of important scientific measurements. Moreover, as the expedition found animals living in the sea that had earlier only been seen on land as fossils, this landmark discovery was a significant step in helping to validate Darwin's theory.

John Murray immodestly but fairly described the consequent report as "the greatest advance in the knowledge of our planet since the celebrated discoveries of the 15th and 16th centuries."[309] The *Challenger*'s journey and the subsequent publications made clear that multidisciplinary study of the oceans was a field of science in its own right. As the first true oceanographic voyage, the Challenger Expedition under the leadership of Charles Wyville Thomson—bringing, as it did, significant new knowledge to the world—laid the groundwork for what would become an extensive and significant academic discipline.[310]

10

William Speirs Bruce (1867–1921)

In all the world there is no desolation more complete than the polar night.
It is a return to the Ice Age—no warmth, no life, no movement.

–Alfred Lansing

The British Imperial Century began with the defeat of Napoleon and ended with the outbreak of the First World War. That 100-year-period marked one of the longest intervals of relative peace that Great Britain and the rest of Europe had experienced, so resources previously earmarked for the build-up of the tools necessary for war could be utilized for exploring voyages that enhanced every aspect of the empire. The men and ships of the Royal Navy—now no longer needed to defeat aggressors, conquer new territory, or establish trade routes—supplied the personnel and equipment necessary to push the exploration ever forward. Expeditions continued to be launched at a rapid pace. But by the final decade of this momentous century the role of the medical explorers had shifted almost exclusively to journeys devoted to the advancement of science. With six of the seven continents now explored, attention turned to the polar regions.

At this point in time the explorers whose activities were filling the newspapers were working in the Arctic and Antarctic. The names of brave adventurers such as Amundsen, Scott, Peary, and Shackleton were constantly in the news as these men attempted to reach the north or south poles, or cross from one side of the seventh continent to the other or establish the boundaries of lands that not too many years earlier had been completely unknown. Roald Amundsen made three trips to the polar regions. He received the Hubbard, Vega, and Daly medals for his efforts, and although Norwegian he was awarded the Congressional Gold Medal by the United States. Robert Falcon Scott made two voyages to Antarctica. Along with a knighthood, he was awarded the Cullum Geographical Medal by the American Geographical Society and the Patron's Medal by the Royal Geographical Society in London. Upon his untimely death, memorials were erected throughout Great Britain in his honor. Ernest Shackleton, who made three trips to Antarctica, was knighted and awarded the Polar Medal. Robert Peary, an American, made five trips to the polar regions; he was awarded the Cullum, Daly, and Hubbard medals.

But within this cohort of individuals, one name remains surprisingly obscure. William Speirs Bruce completed 13 separate polar voyages, of which 2, the first in 1892–93, and the second in 1902–04, were to the Antarctic regions. He also made 11 successful voyages to the Arctic Regions between 1896 and 1915.[311] During his lifetime he received many awards: the Royal Medal from the Royal Society of Edinburgh, the Patrick Neill Medal from the Royal Society of Edinburgh, and the David Livingstone Medal from the American Geographical Society.

But the honors that eluded Bruce were knighthood and the Polar Medal. The latter, awarded by the Sovereign on the recommendation of the Royal Geographical Society, went to individuals for outstanding service to the field of polar research. From 1857 until 1968 it was presented to anyone who participated in a polar expedition endorsed by the government of any Commonwealth country or region; it was awarded to the members of every other British or Commonwealth expedition. Knighthood was bestowed upon the British expedition leaders Scott and Shackleton. Yet although Bruce was nominated repeatedly, both knighthood and the Polar Medal were denied to him.

The road to Edinburgh

William Speirs Bruce was born in London on August 1, 1857. Although born in England, he had Scottish ancestry, and he ultimately evolved into one of the most fervent Scottish nationalists of his time. He was the fourth child of Samuel Noble Bruce, a Scottish physician, and Mary, his Welsh wife. As William was the son of a successful doctor, his background was privileged but not exceptional. His childhood differed, however, from others of his social class in one important respect; the Bruce children were not sent away to boarding school at an early age, and the family did not employ tutors or governesses to educate the youngsters at home. Instead, William's first teacher was his grandfather. The Reverend William Bruce hailed from Glasgow and his wife was from Orkney; each day he tutored William and his seven brothers and sisters in the family's home. When William turned 12 years of age, he was sent to boarding school in Norfolk, where he remained until 1885. He then transferred to University College School in Hampstead, where he prepared for his matriculation examination for admission to the medical school at University College London (UCL). He succeeded in gaining admittance and in the autumn of 1887 was preparing to start his medical studies in London.

During the summer preceding his projected enrollment at UCL, Bruce was offered the opportunity to attend a pair of six-week vacation courses, in botany and oceanography. The botany classes were held at the Edinburgh Botanical Gardens, while the oceanography classes were held at the Scottish Marine Station on the Firth of Forth. Eminent scholars such as the physicist Peter Tait, naturalist John Thompson and anatomist William Turner gave lectures at the university for the summer students. Access to these courses also meant the option of studying under Alexander Buchan, who would become the founder of the British Meteorological Service, and Patrick Geddes, who would be credited as the father of the study of ecology. In the summer of 1887, with the recent success of the Scottish-led Challenger Expedition, and the University of Edinburgh's ascent into its golden period of academic life, there was no better place for a young man with a strong interest in the natural sciences.

Bruce was also allowed to work as a volunteer in the Edinburgh laboratories under Dr. John Murray, where specimens brought back from the Challenger Expedition were being examined and classified. Inspired by all he had seen and done, Bruce determined that he must stay in Scotland, so he gave up his place at UCL and enrolled in the medical school at the University of Edinburgh. His formal medical studies allowed him to remain in contact with the professors from his summer courses and, although he was only a first-year medical student, these esteemed scientists soon became his colleagues.

Bruce studied medicine from 1887 to 1892. (**Figure 10.1**) Records indicate that he was a good student, and he completed courses in philosophy, botany, embryology, anatomy, histology,* zoology, chemistry, physiology, and natural history.[312] His time in Edinburgh was highly successful, and he became associated with eminent scholars who would impact positively on the rest of his career. But the study of medicine could not hold his interest, and he began spending more and more of his time working with John Murray on sorting and cataloguing the findings from the Challenger Expedition. In 1892 he interrupted his medical studies to participate in what would be the first of his expeditions into the polar regions.

Figure 10.1 William Speirs Bruce (Royal Geographical Society)

The Dundee Expedition

By the 1890s it had become apparent that through over-hunting the population of whales in the North Atlantic was in serious decline. Although at the time of Bruce's studies in Edinburgh the market for whale oil and blubber was also declining as gas and electricity replaced oil for heating and lighting, the demand for baleen, the flexible filter plates from the jaws of some whales, was soaring. Baleen was used in much the same way that plastic is used today, and

---

* The branch of biology which studies the microscopic anatomy of biological tissues

when harvested was used, among other products, in the manufacture of umbrella spokes and corset stays. The baleen taken from a single whale could be worth £2,000–£3,000. To address the problem of the whale shortages in the North Atlantic, the Royal Scottish Geographical Society and the Royal Geographical Society of Australia set up the Antarctic Exploration Committee and began looking at the possibility of the Antarctic Ocean to fill this need. The Dundee Whaling Expedition would be the initial attempt at investigating the potential for whaling in these southern waters.

Robert Kinnes of Dundee was charged with fitting out four whaling ships with auxiliary steam engines to undertake this initial exploratory venture. The ships were the *Balaena, Active, Diana,* and *Polar Star,* and among these four ships the expedition included a total crew of 130. As with most exploratory voyages at this time, the

Figure 10.2 Hugh Robert Mill (Royal Geographical Society)

journey included varied goals; in addition to the commercial nature of the voyage, scientific observations and oceanographic research were to be carried out on the ships. Kinnes had been approached by the Royal Geographical Society (RGS) to include naturalists among the ships' crews. As they had done for decades, medical doctors could fill the dual role of surgeon and naturalist, and their inclusion could result in additional funding provided by organizations such as the RGS. Bruce was recommended for the expedition by Hugh Robert Mill, a Scottish geographer and meteorologist who was serving as librarian for the RGS in London. **(Figure 10.2)** Although Bruce had not completed his medical studies, he was listed as surgeon on the *Balaena,* and a good friend of Bruce's, artist William Gordon Burn Murdoch, was designated assistant surgeon.

While Bruce had some medical knowledge, he did not merit the title of surgeon— and Burn Murdoch, having not trained as a doctor, had no medical experience at all. It was clear that Bruce had been included because of his background in natural history and that Burn Murdoch was there because of his artistic abilities; fortunately for them and the crew of the *Balaena,* there were no serious illnesses or injuries for the pair to deal with during the voyage.[313] The RGS provided the equipment for the four vessels under the guidance of veteran Arctic explorer Benjamin Leigh Smith. He chose instruments for geographical observations, including deep-sea thermometers, an aneroid barometer, and a spectroscope, along with detailed directions to assist in the collection and preservation of the natural history specimens.[314]

Now fully staffed and fully equipped, the four ships left Dundee on September 6, 1892. They were able to leave port, but foul weather forced them to remain near the coast of Scotland for over three weeks. On December 8, 1892, after 93 days and 7,000 miles, the *Balaena* and its sister ships finally reached the Falkland Islands. While in port, the officers and surgical team were hosted by the governor and his wife, Sir Roger and Lady Goldsworthy, and through this chance encounter the pair were destined to play a role in Bruce's future. The socialization was brief, however, and after a three-day rest and laying in supplies, the four ships left the Falklands. On December 16 they sighted the first iceberg; Burn Murdoch described it as over a half-mile long and over 200 feet high. They were soon at their appointed location and began the search for what they assumed would be a plentiful supply of whales.

It soon became obvious that there were no whales in the area, and the captains of the four ships decided to cut their losses by filling the ship's holds with skins, oil, and blubber from the seals that were present in abundance. Despite their assigned roles, Bruce and Murdoch were also required, along with the other naturalists and artists, to "work in the boats." Although clearly aboard to make meteorological and other scientific observations, they were obliged to participate in harvesting the seals. Bruce was appalled. The animals had never seen humans and lay passively as the whalers advanced; the indiscriminate killing resulted in the death of over 20,000 seals. But even then, although the four Dundee ships returned to port with their cargoes of seal oil, blubber, and skins the expedition met with a huge financial loss.[315] By May 1893 Bruce was back in Scotland, knowing that the voyage had been a failure not only commercially but also scientifically; no whales had been sighted in the regions explored, and the scientific output from the voyage was, in Bruce's words "a miserable show." [316]

For the naturalist and the artist, the frustration was generated by how completely the commercial goals of the expedition had dominated the scientific ones. Aside from the fact that the two men had been required to take part in the mass slaughter of the seals, it had been clear from the beginning that the ship's captain had no interest in or understanding of the role the naturalists were to have played. Burn Murdoch complained that their scientific work was not just hindered but "mocked and jeered." Bruce reported to Mill at the RGS that he had been denied access to the ship's navigational charts, which made it impossible for him to fix the location of his observations. No facilities had been provided for the preparation of specimens, and as a result many of them were lost due to careless handling by members of the whaling crew. Nevertheless, Bruce's letter to the RGS ends with, "I have to thank the Society for assisting me in what has been, despite all the drawbacks, an instructive and delightful experience." In a letter to Mill, Bruce outlined his wishes to travel again by saying, "the taste I have had has made me ravenous." [317]

Despite the limited scientific success of the expedition, both Burn Murdoch and Bruce were able to publish several reports and articles. Great interest was shown in the expedition throughout Scotland and England, and Bruce was soon elected as an Ordinary Fellow of the Royal Physical Society of Edinburgh, an organization founded "for the cultivation of the physical sciences." He started to dream about leading his own polar expedition, and without hesitation, and without any discussion with his father, he made the decision to drastically alter his career path. With less than a year to go to complete his medical studies, he abandoned medicine and became a naturalist. Why he made that decision at that time is open to speculation; perhaps, having experienced the possibilities of scientific discovery in the South Atlantic, he decided that research was more suited to his abilities. Bruce has been described as "single-minded," "serious," and "difficult to divert." [318] That being so, having come to this momentous conclusion, he probably saw that there was no point in delaying his decision.

Ben Nevis and Franz Josef Land

Bruce had returned to Edinburgh with no money and little prospect of employment. He must have reflected upon his recent decision to abandon his medical studies; as a doctor he would have had a steady income that would have at least partially supported his newly developed interest in polar exploration. The decision to leave medicine had devastated his relationship with his father, who had as a result cut off his financial support. Bruce would have to seek funding for his research from other sources. He wanted to return to Antarctica, where he saw the greatest potential for scientific study, and he began searching for financial support among a wide variety of individuals and organizations.

As distressed as Bruce had been with his forced participation in the slaughter of the seals, he understood that the best option for supporting his scientific studies was by incorporating his work within a commercial effort. He began to contact companies throughout Europe who were involved in whaling. Receiving little or no support from the whalers, he next contacted Sir Roger and Lady Goldsworthy, encouraging them to consider support for various agricultural projects that would be located in the Falklands. Although Sir Roger was still Governor of the Falkland Islands, the Goldsworthies, too, expressed little interest in supporting the agricultural endeavors or Bruce's ongoing research.

Having met with no success in these commercial appeals, Bruce turned to the scientific community and approached his contacts at the RGS and the British Museum. This resulted in a request for him to make presentations to the British Association for the Advancement of Science* and to the Royal Physical Society of Edinburgh. But neither the commercial approach nor his appeals to the scientific community proved successful in providing additional funds, and it seemed that nothing was going to materialize from his efforts.

By this time, his financial state had reached a point of desperation, and in 1895 he applied for the vacant post of Curator of the Raffles Library and Museum in Singapore. He wrote that he was 28 years old, unmarried, and in perfect health. He listed his university course work and recent expedition experience. This application was supported by 15 letters of reference provided by some of the top scientists in Scotland and England.[319] But, in large part because he did not have a university degree, he failed; the post was awarded to another candidate. Although it must have seemed at the time that this was a final setback, it actually set the direction for the remainder of his career.

---

* Now the British Science Association

The High-Level Meteorological Observatory had been officially opened on October 17, 1883. It is on the summit of Ben Nevis, the highest mountain in the United Kingdom, which stands at the western end of the Grampian Mountains in the Scottish Highlands. Shortly after being rejected by Raffles, Bruce applied to replace Robert Traill Omond, who had been superintendent of the observatory since its opening. Although that position was given to Angus Rankin, and Bruce accepted the role of Rankin's second assistant, for the rest of his life Bruce would indicate that he had been appointed superintendent of the meteorological station as Omond's replacement.

Regardless of Bruce's title, his time at Ben Nevis was an important period for him, and the skills he developed were ideal preparation for what would be needed during his future polar explorations. His method of obtaining readings from both the summit and the base of the mountain was of great value in understanding weather patterns at various elevations, and was extremely attractive to the scientific community. In addition to providing him with some much-needed financial support, his Ben Nevis work became almost an apprenticeship in scientific investigation and scientific discipline. The weather conditions there were generally extreme, and during the winter months he could only get to the external instruments through a series of snow tunnels designed to provide access from the observatory building to the outside world. He became adept at meteorological observations and compiling and recording scientific data. He corresponded and met with other scientists, and developed relationships which would benefit him in the future. But not all of his associations were newly formed, and in June 1896, he received a telegram from his long-time supporter, Hugh Robert Mill, at the RGS. Mill had been asked to suggest a naturalist to join the Jackson–Harmsworth Expedition, which was entering its third year on Franz Josef Land.* If Bruce could be prepared to leave in five days the job was his. He accepted the offer immediately.

The stated purpose of the Jackson–Harmsworth Expedition was twofold. Running from 1894 to 1897, it had been charged with carrying out scientific studies and, on the ship *Windward*, searching for a route to the North Pole. But these stated goals had little to do with reality. George Frederick Jackson was a noted big game hunter who cared little about the declared objectives of the expedition. His primary motivation was the personal fame his participation would bring him, along with an opportunity to collect as many hunting trophies as possible during his three years in the Arctic. Alfred Harmsworth, the financial backer of the expedition, was a man of significant influence. He was a pioneer newspaper proprietor who had founded the *Daily Mail* and the *Daily Mirror* newspapers in London, and between his newspapers and other periodicals his publications reached a daily distribution of nearly a million copies. The publishing industry had learned through Stanley's "discovery" of Livingstone that coverage of these expeditions sold newspapers, and Harmsworth felt that his sponsorship of the expedition was a guaranteed method of increasing readership.

Bruce arrived in London in June 1886, less than eight hours of the *Windward*'s departure for Franz Josef Land. He would later write that he worked best under pressure. And Bruce had the knack of inducing pressure on himself. When asked about his luggage, he admitted

---

* A Russian archipelago in the Arctic Ocean

that he had not brought any, so he was forced to borrow items of clothing from the other crew members and from friends who had come to see him off. The same was true of his scientific equipment; even though he was aware that he would be serving in a scientific capacity for at least a year, he had purchased nothing in advance of the trip and was forced to use equipment provided by other members of the expedition or to make implements from the supplies he could salvage upon arrival at Franz Josef Land.[320]

Jackson took three additional and well-respected scientists along on this leg of the tour: the botanist Henry Smith Fisher; Albert Armitage, in charge of astronomical work; and Reginald Koettlitz, a geologist. With these esteemed colleagues, and despite his lack of preparation, Bruce wrote that he greatly anticipated the opportunities for extensive research into the natural history of Franz Josef Land without the frustrations that he had encountered on the Dundee Expedition. The voyage proceeded without incident, and the party soon arrived at the previously established base at Cape Flora.

There were plenty of opportunities for observations and data collection, and despite his hurried departure from London and his shortage of specialized collecting equipment, Bruce managed to collect a significant amount of data, record characteristics of the varied animal life, and bring back an extensive array of eggs, skeletons, and skins. But on his return, he complained that he had been assigned superfluous tasks, including an obligation to skin the bears, birds, and other animals that Jackson had shot. Bruce was opposed to hunting for sport and did not own a gun. But Jackson, who actually made his living as a big game hunter, shot over 100 polar bears during his three years in Franz Josef Land, as well as seals, walrus, foxes, and birds of all kinds. The preservation of these trophies fell to a very discontented Bruce.

In 1897 the expedition ended without any attempt to search for a route to the North Pole. However, two events occurred during Bruce's final days on Franz Josef Land that would significantly impact his life and career. The first was an unexpected visit by the *Balaena*, Bruce's ship from the Dundee Expedition. The ship was now under the command of Thomas Robertson, who had earlier captained the *Active*. With this captain, unlike the previous one, Bruce had a strong professional relationship. This renewal of their acquaintance led directly to the recruitment of Robertson as sailing master for what would become Bruce's most famous voyage. The second event occurred after the *Windward* had arrived in July 1897 to take the group home; a chance variation in course on the homeward voyage took the ship past Bear Island.* This Arctic breeding ground, with its polar bears, arctic foxes, reindeer, and wide variety of sea birds, immediately attracted Bruce, and in later years Svalbard would capture much of his time and attention.[321]

Creating contacts

After the Jackson–Harmsworth Expedition, Bruce went back to Edinburgh. As he was once again without employment and with no private income, he worked briefly at his old job at Ben Nevis, but that turned out to be short-lived. Fortunately in 1898 he was offered the opportunity

---

* Bjørnøya, the southernmost island of the Svalbard Archipelago, in the western portion of the Barents Sea

to return to the Arctic with Major Andrew Coats, the head of a wealthy manufacturing family, who was planning to travel on his steam yacht *Blencathra* on a hunting and sporting trip to Spitsbergen.* Coats had invited Hugh Robert Mill to join him, but when the RGS refused to grant him, their librarian, leave, he suggested to Coats that Bruce be invited in his place. This invitation would prove to be a turning point in Bruce's life. Prior to this, his expeditions had returned moderate results at best. But that was about to change.

The *Blencathra* left port on May 1, 1898, bound for the Arctic. Once again Bruce was unprepared for the journey, and this time he was forced to stay behind for the delivery of a new piece of equipment. When it arrived, he was compelled to travel on a mail steamer, and was not able to catch up with the *Blencathra* until the ship had reached Tromsø, in northern Norway. The hunting party soon departed for Spitsbergen but found that even though it was now midsummer the way was blocked by pack ice. There was no option to go forward, and the party returned to Tromsø, where Bruce saw the *Princess Alice* for the first time. The beautiful sailing ship, which belonged to Prince Albert of Monaco, was said to be the best-equipped ship for oceanographic work since Thomson's Challenger Expedition. Prince Albert, although by interest rather than training, was considered among the leading oceanographers of his time. Coats introduced Bruce to the prince, who invited him to accompany him on a hydrographic survey of the Svalbard Archipelago in advance of a more extensive exploration to be conducted in 1899. This was an unexpected and extraordinary opportunity for a relatively junior scientist, especially as Bruce knew he would be in the company of some of the best oceanographers in Europe.

As the *Princess Alice* was better suited to navigate the ice than the *Blencathra*, they were able to leave immediately. The voyage initially took them to Bear Island, where Bruce climbed Mount Misery and collected birds and plants, before the ship sailed for the west coast of Spitsbergen. Bruce assisted with the hydrographic survey of the waters around this large island and also conducted regular meteorological observations and collected data on variations in the temperature of the sea's surface. During the latter stages of the voyage he was placed in charge of the voyage's scientific observations.

At the end of the expedition he returned to Edinburgh. As he had been given the opportunity to carry out work as a scientist on this superbly equipped survey ship, he had achieved a great deal on the voyage. With the help of the experts aboard the *Princess Alice* he had been able to build on the knowledge he had developed on previous voyages and his work on Ben Nevis.[322] Shortly after his return to Scotland, he received an invitation from Prince Albert to winter in the palace in Monaco; he accepted, and during the winter he sorted and catalogued his Arctic specimens. In the summer of 1899, he again traveled with the prince to the Arctic.

After the second voyage Bruce published his account of the expeditions in the *Scottish Geographical Magazine*. He was now considered one of Britain's most experienced polar explorers; he wrote papers, gave talks, and became more and more confident of his worth and the value of his work. But then he began openly criticizing other explorers, seemingly

---

* The largest island of the Svalbard Archipelago

unaware that his comments could be counterproductive. The RGS, and Hugh Robert Mill in particular, had championed him often in the past. But even the RGS was not immune to Bruce's criticism. When the society agreed to publish one of his reports, he wrote bluntly, "I am not inclined to conjure the confidence of the RGS or any other society. My work must stand on its own merits." [323] What Bruce apparently viewed as confidence in his work came across as a prickly attitude and an unwillingness to compromise. He would learn that making enemies among the highest echelon of one's chosen profession may not be the best avenue for success.

The competition begins

Toward the end of the 19th century great interest was being shown in Antarctica. This was true not only in Great Britain, but also throughout Europe and the United States. Information was needed to determine if this was the seventh continent, because although numerous land sightings had been made it was not yet known if these were isolated islands or part of an actual continent.

In addition, an understanding of the Earth's magnetism was vital to navigation, but a complete understanding did not exist at that time. Without an exact location for the south magnetic pole, precise navigation in the Antarctic Ocean held a great deal of uncertainty. That location was needed for exploration, commerce, and the safety of the ships. The need for further exploration was not in dispute—but the many questions regarding who should fund these voyages, who would control them, and who would select the personnel to be involved were far from settled. Should power be in the hands of the government, or of private businesses, or should it be ceded to scientific societies?

In November 1893, after decades serving as the RGS secretary, Sir Clements Markham became its president and declared that Antarctic exploration would be the cornerstone of his presidency. In a presentation he said that British science, "needed a steady, continuous, laborious and systematic exploration of the whole southern region." Then Sir John Murray, Bruce's mentor from his days spent cataloguing the findings of the Challenger Expedition, presented a paper "setting forth the argument for renewal of Antarctic discovery with telling force." [324] Bruce, who was in the audience, was in complete agreement that this effort should reflect a British national character, and immediately volunteered to be part of the scientific team. Shortly after Murray's presentation, Bruce was invited to present a paper to the Royal Scottish Geographical Society (RSGS) on the benefits of a national Antarctic expedition. His presentation was well received, and following this speech the RSGS both promised its full support to the further exploration of Antarctica and backed Bruce's requests to join the British expedition.

Three issues faced the planners: who should fund these voyages, who would be in control, and who would select the personnel to be involved. By far the most problematic was the funding. For the first five years of his presidency, Markham spent the majority of his time working to secure financing for the mission. He gave talks, sought out wealthy individuals, approached governmental officials, and canvassed the various scientific organizations.

Finally, in 1898 the Royal Society and the RGS set up a joint Antarctic Committee based in London, to oversee a project called the Discovery Expedition. Markham would chair the committee and Sir John Murray would be a member. But from the outset, unfortunately, they openly and fundamentally disagreed about the primary purpose of the expedition; it became apparent that Markham, as a geographer, wanted to emphasize geographical exploration whereas Murray wanted the mission to be based on systematic scientific research related to the geology, plants, and animals of the region. Hugh Robert Mill, serving as Secretary of the RGS, reporting on the committee's deliberations, wrote that the work resulted in "a confusion of jealousies, arguments, and misunderstandings." Murray soon began to withdraw from participation in the committee's work and set his sights on forming a separate Scottish expedition.[325]

Worse, the conflict between Markham and Murray quickly evolved into animosity between Markham and Bruce. The relationship fractured in 1899, when Bruce, having informally volunteered for the expedition five years earlier, now formally applied to join it. He received a prompt reply in which Markham stated that no personnel decisions had been made. Then in March 1900, almost a year later, Bruce, having received no additional correspondence from the RGS, wrote again, ending his note with a paragraph that would permanently sour the relationship. He wrote that he was "not without hopes of being able to raise sufficient capital whereby I could take out a second British ship to explore the Antarctic regions."[326] He was apparently presenting this as a collaborative and complementary proposal to the Discovery Expedition. But Markham clearly did not see it in the same way, and his response was quick and angry:

> I am very sorry to hear that an attempt is being made at Edinburgh to divert funds from the Antarctic expedition in order to get up a rival enterprise … I do not understand why this mischievous rivalry should have been started, but I trust you will not connect yourself with it.[327]

Markham's response is easy to understand. He had been attempting to raise funds for the Discovery Expedition for five years, and clearly saw the Scottish efforts as syphoning funds that could or should have gone to him. Murray and Markham already differed on what each saw as the primary focus of the mission, and the tone of Markham's letter now resulted in the Edinburgh contingent making the decision to go their separate way. Sir John Murray, Chair of the RSGS, announced plans for a Scottish expedition to the Weddell Sea,* and stated that the exploring effort would be led by Bruce. Sometime later, Bruce and Markham finally met, and Bruce was offered a position of naturalist on the *Discovery*. But this was too little too late; Bruce responded, "No thank you. I have arranged an expedition of my own." [328]

The Scotia Expedition

In 1901, a year before the Scottish expedition would occur, Bruce had married Jessie Mackenzie. She, having trained as a nurse, had met her future husband while working in his

---

* Part of the Antarctic Ocean, next to the northernmost part of Antarctica

father's medical clinic in London. At first the marriage seemed a happy one and in April 1902, Jessie gave birth to a son who they named Eillium, a Scottish Gaelic equivalent of William. But as Bruce had already begun to prepare for the expedition he was frequently away from his wife and son. Planning for the voyage consumed all his time, and the dogged focus that he employed in his work sadly would eventually destroy the marriage.

His first task was to acquire a ship, and although he initially considered purchasing the *Balaena*, the vessel from his days on the Dundee Expedition, it was far too costly. He settled on the *Hekla*, a Norwegian whaler that he purchased for £2,620. This ship, which Bruce rechristened *Scotia*, appeared to be in good condition. But the purchase had been hasty. Upon closer inspection when it had been brought to Scotland, Bruce discovered that it needed a complete overhaul. The refit increased the cost to £16,730, but the end result was an incorporation of all the improvements the crew would need for their work in the Antarctic. However, the added costs meant that if the voyage was to proceed, additional fundraising would be required. Bruce again turned to Coats, who ultimately contributed £30,000 out of the total cost of approximately £37,000.[329]

Next Bruce began recruiting personnel for the voyage. Thirty-three men were selected to serve as crewmembers on the *Scotia*, and it is evident from the ship's log that Bruce concentrated on recruiting almost exclusively from within the Scottish borders. He personally selected the scientific staff, again Scots, who were all friends or colleagues that he had worked with in the field. Captain Thomas Robertson, who Bruce had known since the Dundee Expedition, would serve as sailing master on the voyage. The expedition, renamed the Scottish National Antarctic Expedition, was now ready to set sail, and the expedition members were given a farewell dinner hosted by Sir John Murray on behalf of the RSGS. On November 2, 1902, the *Scotia* departed from Troon, on the north-west coast near Glasgow. If by naming the voyage the Scottish National Antarctic Expedition, by employing an almost exclusively Scottish crew and scientific staff, and by honoring the ship in giving it the original name of Scotland left any doubts as to the nationalistic nature of the voyage, those doubts would be laid to rest on seeing it departing from the west coast of Scotland flying not the Union Flag but the blue and white St. Andrew's Cross.

The *Scotia* sailed south, stopping at Madeira and Cape Verde before reaching the Falkland Islands on January 6, 1903. Sir Roger Goldsworthy, who had died in 1900, had been replaced as governor by Sir William Grey-Wilson, but Bruce and the crew of the *Scotia* were well received, and they spent the next three weeks at Port Stanley taking on supplies. By February 3 the ship had reached the South Orkney Islands, located at the northern edge of the Weddell Sea. Specimens were collected, dredging was begun, and depth soundings and hourly meteorological readings were taken. But the summer season was ending, and it was important to locate a sheltered area to overwinter. Captain Robertson settled on a bay on the south side of Laurie Island. **(Figure 10.4)** The ship was allowed to become frozen into the ice, and Bruce renamed the location Scotia Bay.

The ship would remain at that location until freed by the spring thaw in November 1903. Throughout that period the members of the scientific team conducted extensive research. They took more meteorological readings, trawled for marine samples, and collected biological

Figure 10.4 Piper and Penguin (Wikimedia Commons/PD)

and geological specimens. They also made frequent trips from the *Scotia* to the beach, where they constructed what would become a permanent station to be used by scientists on future expeditions. Bruce named the station Omond House after Robert Traill Omond, the director of the Edinburgh Observatory. **(Figure 10.5)** Although the principal building is now a ruin, it has been designated the oldest scientific station in the Antarctic, and with new facilities the site has remained in continuous use since its construction by the scientists from the *Scotia*.

Once freed from the ice in the spring the *Scotia* made for open water, and after a short stop at Port Stanley it set sail for Buenos Aires. Upon arrival Bruce was able to report to the British Consul on a highly successful oceanographic voyage of 4,000 miles across hitherto unexplored ocean and the construction of the first permanent scientific station in Antarctica. Bruce immediately pursued plans for the future of the new scientific base. He was hopeful that the work at Omond House would continue when the *Scotia* returned to Scotland, but it turned out that neither the Admiralty nor the Foreign Office had any interest in the South Orkney Islands. Both bodies encouraged Bruce to transfer the station's ownership to Argentina. He complied, and the transfer was designated as a "continuation of the meteorological, magnetic, and biological work on Laurie Island." [330]

The *Scotia* was refit in Buenos Aires, and then returned to Laurie Island to land provisions and retrieve four members of the party who had remained there. The ship continued exploration of the Weddell Sea, and then sailed for Cape Town. It then continued north, calling at St. Helena and the Ascension Islands before finally reaching Glasgow on July 21, 1904. The expedition team had fulfilled all scientific expectations and brought back an extensive assortment of specimens. Their work was historic as it changed the way the geography of Antarctica was understood. New land discovered in the Weddell Sea proved that the coast of the continent extended 500 miles farther north than had been previously believed. From the thousands of soundings and meteorological recordings it was discovered that the behavior of sea ice in the Weddell Sea controlled the weather patterns in South America. Furthermore, 1,100 types of animals were catalogued— including seven new genera and more than 24 hitherto undiscovered species.[331] What Bruce could not have known was that at just 37 years of age he had just reached the apex of his scientific career.

*Figure 10.5 Omond House
(Wikimedia Commons/PD)*

The beginning of the end

Bruce's accomplishments were acclaimed by the scientific communities in both Scotland and England, and he was awarded the Gold Medal of the Royal Scottish Geographical Society. But recognition of the expedition's success was not forthcoming by the general public, nor were Bruce or any of his company awarded the prestigious Polar Medal by the RGS. This snub only heightened his Scottish nationalism, and his feelings that the RGS was hostile to any achievement "north of their border." [332]

He continued to be hounded by his creditors. He would have preferred to stay in the Antarctic for a second season, but in spite of the generosity of the Coats family his finances were only sufficient for one year of exploration. Finally, his plans to retain the *Scotia* for future voyages were dashed when he was forced to sell it to cover his debts from the voyage just completed. Shortly after his return, he developed a prolonged attack of influenza, forcing the account of the expedition to be written by three of his colleagues. Faced with no prospects of a quick return to the polar regions, and having delegated the journal summary to others, he threw all of his energy into what he would call the Scottish Oceanographical Laboratory.

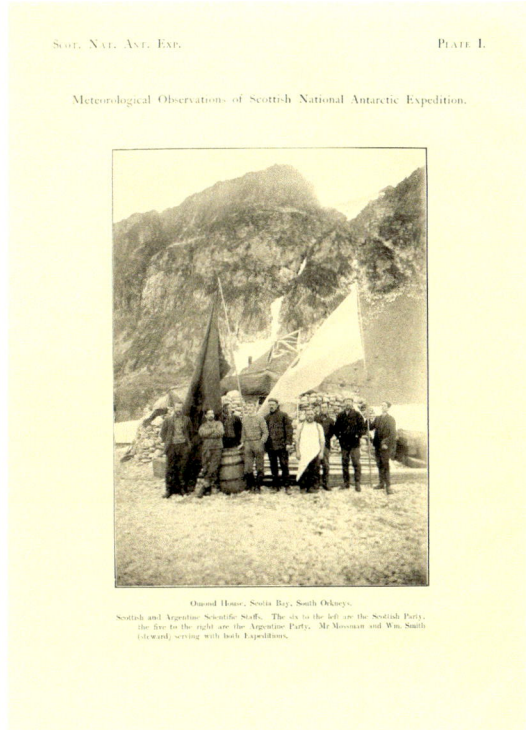

Through his previous Arctic and Antarctic endeavors, Bruce had amassed a huge collection of biological, zoological, and geological specimens, as well as detailed records of the data that had been collected, so he needed a museum/laboratory to house his extensive collection. A single-story building became available near Surgeon's Hall in Edinburgh, and he promptly purchased the structure. It was formally opened by His Serene Highness Prince Albert I of Monaco in 1906 at a ceremony attended by the key members of Edinburgh's civic and academic communities. Over the years of its existence, the Scottish Oceanographical Laboratory became a hive of activity; in the laboratory, the Scotia Expedition's reports were produced, and a public space allowed visitors from England, Europe, the United States, and the Far East to examine the collection. The center also played host to some of the most noted explorers of the age, including Ernest Shackleton, Fridtjof Nansen, and Roald Amundsen.[333]

Unfortunately, even with the funds—£5,000—raised through the sale of the *Scotia*, Bruce still owed money for the expedition's costs and was forced to continue a frustrating hand-to-mouth financial struggle. He did not mind the poverty but seemed oblivious to the effects it was having on his family. This single-minded involvement in his projects, and the lack of attention he paid his wife and son, had disastrous consequences, and the domestic situation was not helped when in 1909 their daughter Sheila Mackenzie was born; a fourth mouth to feed. The tension between the couple mounted as gradually Jessie had to shoulder all responsibilities for the family's financial needs and was forced to rely on help from friends to feed her family.

Yet Bruce continued to make his work his primary focus. He knew that if he hoped ever to return to polar exploration, he would need to combine larger commercial endeavors, such as agriculture or mining, with his scientific mission, and during the next decade he attempted to launch a variety of schemes, many of these involving returns to Spitsbergen and the Barents Sea. But each one took him away from home for a prolonged period of time, and none produced anything close to financial success. On his return to Scotland in 1915, he and Jessie would no longer live as husband and wife—yet although the marriage was all but over they continued to live in the same house.[334]

Bruce always hoped that the Scottish Oceanographical Laboratory would become the foundation for a Scottish National Oceanographical Institute to be housed in a purpose-built building on the University of Edinburgh's campus. But by 1919 he had been unable to raise sufficient funding for the laboratory, and he was forced to close the facility and disperse the collection to a variety of universities, museums, and societies.[335] Few remembered his achievements from the Scotia Expedition. His health was beginning to fail, and he was estranged from his wife. For his final 18 months, Bruce was repeatedly in and out of hospital and on October 28, 1921, he died.

Even in Scotland William Speirs Bruce has been all but forgotten. His fierce defense of Scottish nationalism, his substantial ego, and his determination to push exploration for the sake of science and geographic discovery when the rest of the world was looking for heroic endeavors alienated him from both the general public and the scientific community. No plaque or statue has ever been erected in his memory. His plan for a Scottish oceanographical institute was short-lived, and eventually came to nothing. The various commercial endeavors he launched to support his ongoing scientific efforts never turned a profit. Neither he nor any of his expedition's members were ever honored with the Polar Medal from the RGS in London.[336] He never seemed to understand that his fervent Scottish nationalism, coupled with the belief that England in general and the RGS in particular were his enemies, had driven potential allies away from his camp.

He had, however, clearly learned that his philosophy of exploration solely for the sake of science—at least within the polar regions, where the goal was to travel the farthest or fastest—was no longer the currency of the realm. Although he never wavered from his approach, searching simply for the sake of science, and he continued to push for the public to acknowledge his type of exploration, there was clearly an underlying knowledge that his methodology would never capture the public's attention. He had come to realize that in Edwardian Britain it was Scott and Shackleton who epitomized the heroic figure, as shown in Bruce's 1911 book, *Polar Exploration*, the only book he created for the general public:

> The world shrinks and now there are few parts of the globe which have not been traversed. I say purposely traversed, for many parts traversed have not been explored … What the mass of the public desire is pure sensationalism, therefore, the Polar explorer who attains the highest latitude and who has the powers of making a vivid picture of the difficulties and hardships involved will be regarded popularly as the hero … while he who plods an unknown tract of land or sea and works there in systematic and monographic style will probably never have such worldly success.[337]

11

Conclusion

Wherever you look, there are still things we don't know about and don't understand.
There are always new things to find out if you go looking for them.

–David Attenborough

There is no doubt that during the British Imperial Century it was the Scots who led Great Britain in explorations of the globe. The disproportionate numbers of Scotsmen electing to become explorers, when compared to their English counterparts, is easily quantifiable. It cannot be refuted that a significant number of the explorers were doctors who had been prepared in Scottish universities; the extensive list can be found in the appendix. It is also without dispute that in 1707 England and Scotland became the Kingdom of Great Britain; that between 1740 and 1820 the Scottish Enlightenment occurred; and that during the British Imperial Century the Scottish universities were creating a radically different type of doctor than the English universities. These are all facts, but this book offers a theory—a description of the facts that explains why something occurs. In this case the theory is that the creation of the Kingdom of Great Britain, combined with the Scottish Enlightenment, Edinburgh's introduction of a radically new methodology in preparing physicians and surgeons, and the changing nature of exploration during the British Imperial Century created the medical explorers of Edinburgh. An examination of the interconnectedness of these individual events begins to illuminate the theory.

Opportunities in the joint-stock companies, the Royal Navy, and in the various organizations involved directly in exploration, such as the Royal Geographical Society and the Royal Botanic Gardens Kew, became available to the men of Scotland because the Acts of Union, uniting their kingdom with England, provided them with a partnership which greatly expanded their employment opportunities. These opportunities had long been available to Englishmen, and organizations such as the East India Company and the British Military were the almost exclusive domain of the English until 1707, when the Acts of Union opened these options, and others, to Scotsmen. Two aspects of Scottish culture which greatly impacted these men were their close relationship with the sea and their willingness to move beyond their nation's borders. These were the sons of sailors and the sons of crofters. Possessing a strong character, a willingness to accept hardship, and a tradition of travel in search of a better life, many accepted these employment opportunities without hesitation. Soon both the joint-stock companies and the British Military were disproportionately populated by the Scots.

The Enlightenment produced a more sophisticated and educated Scottish society which allowed men from the middle and working classes to become physicians and surgeons. But without the changes in the way Edinburgh prepared its medical graduates those newly trained doctors would not have been any better prepared to lead Britain's exploring efforts than those graduating from English or Irish universities. It was through the Scottish government's strong commitment to "universal" education,* and a university admission system that was not influenced by social class or religion, that Scotland was preparing doctors who were accustomed to the hardships and lifestyle associated with the working and middle classes. These men were not only better prepared to perform the duties required of the medical explorers but were also much more willing to accept these challenges than the university graduates emerging from England's upper classes.

Three factors—vastly increased employment options, tradition-shattering medical education, and a massive intellectual movement—all occurred in the 100 years preceding the British Imperial Century. It could be said that if just one or even two of these events had occurred the impact would not have been great enough to create this phenomenal group of physicians. But taken in total, with each factor building upon earlier experiences, an exceptional, indeed unique, group of explorers did emerge. It could also be argued that had these events occurred within a country that did not share Scotland's cultural and social identity they would not have had the same impact.

It has been suggested that the British Imperial Century required a new breed of explorer. Research shows that from the beginning to the end of the Imperial Century, the specific charge given to these explorers changed several times. Their exclusive academic background and preparation, their cultural heritage, and the historical events preceding them allowed them to adapt as the reasons for exploring changed. During the course of the British Imperial Century, the medical explorers were called upon to fill a variety of exploring roles: geographical, naturalist, proselytizing and the eradication of slavery, and the purely scientific. Some of these efforts, such as geographical exploration, extended throughout the 100-year period. Others, such as exploring for the sake of science, had specifically defined periods.

Commercialism and the Royal Navy

Although the focus of exploration changed repeatedly, two aspects remained constant. First was the active partnership between the explorers and the Royal Navy. At the end of the Napoleonic Wars in 1815, Britain was firmly established, along with France, Russia, the Ottoman Empire, and China, as one of the world's great imperial powers. But among this elite group, Britain was clearly dominant. Exploration of the far-flung empire required travel

---

* Only 50%, of course. Until the late 19th century girls' education consisted almost entirely of basic domestic skills.

by sea, and now with no naval wars to fight, and a surplus of both men and equipment, the Royal Navy began to provide ships and personnel for exploration of the expanding empire. The medical explorers were well positioned to serve the dual roles of medical officer and explorer. Not only had their robust and expansive academic training given them the necessary intellectual tools but also, possessing a deeper academic knowledge than any of the other ship's officers, they could fill the role of chronicler. Medical explorers were often charged with creating publications highlighting the attributes of the newly explored distant corners of the empire. This was not only important to the government in London, but also essential for instilling nationalistic pride and support among the general population.

The second constant was the British Imperial Century's compulsion for trade. Commercial exploration was not confined to a particular time within the Imperial Century but rather, being embedded within each of the reasons for exploration, it was interwoven throughout the entire period. British merchants, facing few competitors, looked to exploration to advance their trade under the protection provided by the mighty Royal Navy.[338] Alongside the formal control that Britain exerted over its own colonies, its predominant position in world trade meant that it soon effectively controlled the economies of several of the other imperial powers, including those of China and the Ottoman Empire. Explorers were needed to push farther into these unknown lands and to look for new trade routes that might enhance Britain's already strong commercial position, and their ranks included John Rae looking for the missing link for the Northwest Passage. Although he may be best known for discovering the fate of the lost Franklin Expedition, or for mapping large sections of the previously unknown Arctic, both the British Government and the Hudson's Bay Company had in fact directed him to find the Northwest Passage, which would greatly enhance trade between Britain and Asia. Rae, and other medical explorers, were needed to develop resources that could benefit the empire, to establish trading centers, and to install governmental systems that would provide ultimate control from London.[339] The British Government, using the Royal Navy as its primary tool, was more than willing to offer support to geographic exploration or to assist the naturalists. But its primary motive was almost always financial gain.

The changing role of the medical explorers

The first role assigned to the medical explorers was geographic exploration, and it would continue throughout the next 100 years. At the start of the British Imperial Century the British possessions in almost every quarter of the empire were mere streaks along a coastline, or islands adjacent to larger sections of a claimed mainland, and explorers were needed to create maps and charts of the newly acquired territories, or to find usable routes into a continent's interior. Mungo Park, directed to determine the origin, course, and mouth of the Niger River, was among the early medical explorers employed by the African Association. Although Park is an example of a geographic explorer from the beginning of this period, even then commercialism was a primary motive, as his employers wanted to bypass the African middlemen.

During the next 100 years Britain continued to move forward, building its empire. Its efforts were extremely successful; nearly 10 million square miles of territory and roughly

400 million people were added to the realm. Medical explorers were instrumental in surveying and documenting their discoveries in Africa, Asia, Australia, and North and South America. By the end of the century most of those regions had been thoroughly explored, and the need for additional geographic exploration had been pushed to the far regions of the Earth. But the need for geographic exploration had not been fully met; the emphasis was simply moved to new frontiers. So it was that William Speirs Bruce would make his mark by verifying the boundaries of Antarctica at almost the end of the Imperial Century.

The medical explorers also served as naturalists. The efforts of the naturalists involved by far the largest number of medical explorers. From the beginning of the Scottish Enlightenment through to the middle of the 19th century, the British enthusiasm for natural history was abiding and fervent. The interest was reflected in an increased popularity in the museums, zoos, aquariums, and botanic gardens which were soon constructed in most major cities. These institutions began to employ medical explorers who would travel the globe seeking to identify new plants and animals. Archibald Menzies, for example, spent years traveling around the world identifying species hitherto unknown to Great Britain. These efforts not only increased scientific knowledge but also promoted the expanding empire to the general public. These journeys were also designed to identify plants and animals which could provide financial value to the home country, and this undertaking soon evolved into what became known as economic botany. The medical explorers assisted the efforts of promoters such as Joseph Banks in establishing botanic gardens throughout the empire; the creation of his hub and spokes system helped to make Great Britain the greatest economic power of its day.

The medical explorers were for the most part religious men who were opposed to slavery. However, the exploration designed to attain this goal involved by far the smallest number of this group, and their efforts, unlike those of the geographic and commercial explorers, did not extend through the entire period. Even the admirable goal of ending slavery had at its base a commercial aspect, and by the middle of the 19th century a small group of the medical explorers were actively involved in the eradication of slavery and the spread of Christianity. Scotsmen such as John Kirk and David Livingstone believed strongly that the successful commercial exploration which had been developed must go hand in hand with ending the trade in human traffic and the conversion of the Muslims and other non-believers to the Christian faith. By the end of the Imperial Century, this had been accomplished. Slavery in East Africa had been ended, and the missionary work initially conducted by the medical explorers had been subsumed by various religious organizations.

At the end of the British Imperial Century, the role to be played by the medical explorers would change one last time, and they would be charged with their final mission within this 100-year span. The final role for the medical explorers would involve missions conducted strictly for the sake of science. The geographic exploration of the polar regions, heavily influenced by the work of William Speirs Bruce, now turned to voyages designed to assess the unique plants and animals found in those regions. Also, in the same way that botanic gardens had been established throughout the empire, Britain now looked to establish permanent scientific stations within those remote regions.

When exploration had been deemed successful throughout the seven continents, the medical explorers turned to the sea. Led by Charles Wyville Thomson, the first extensive study of the world's oceans began. For three years the crew of the *Challenger* examined all aspects, from the waves on the surface to the deepest trenches on the ocean floor. This would be the final role that the medical explorers would be asked to play. For 100 years they had led Great Britain in exploring the expanding empire. The reasons for exploration had changed repeatedly throughout this period and with each change the medical explorers had successfully adapted. But, although they could not know it at the time, the end of this era of exploration was rapidly approaching, and the role they had been destined to play would soon cease to exist.

Disappearing one by one

The British Empire was at its peak in 1900, when it covered one quarter of the earth's land surface and included roughly one quarter of the world's population. At that time, the empire included the United Kingdom, Canada, Australia, New Zealand, India, Pakistan, Burma, Ceylon, Aden, South Africa, Kenya, Uganda, Tanganyika, the Sudan, Ghana, Nigeria, Sierra Leone, the Gambia, the Ivory Coast, the Gold Coast, Rhodesia, Malawi, Zambia, Zimbabwe, Botswana, and Namibia. Smaller territories included Hong Kong, Singapore, Brunei, Malta, Cyprus, Gibraltar, and the Falkland Islands. The empire also included a number of protectorates, mandates, and colonies, such as the protectorate of Northern Rhodesia and the colony of Southern Rhodesia.[340] Though the methods of colonial administration varied greatly from region to region, each country or territory had a solid system of leadership in place, albeit under the control of a central agency in London. The boundaries of the various spheres of influence had been firmly established, and the routes and methods of transportation within each country were operational. With the empire at its maximum size, and the regions within its boundaries explored, the need for geographic exploration ceased to exist.

The era of the naturalist explorer had seen an attempt to quantify, categorize, and explain the world, and it was the medical explorers of Edinburgh who had emerged as the individuals best suited to perform these tasks. During the early part of the British Imperial Century collecting, by both the general public and the naturalist explorers, became almost an obsession. It had been assumed that there was a finite number of plants and animals in the world, and the overarching goal of the naturalists was to collect at least one of each variety. The Royal Navy had surplus ships, and Edinburgh surgeons aboard these vessels could serve as naturalists in addition to their medical duties. However, in 1859, when Charles Darwin published *On the Origin of Species*, the realization that nature was constantly evolving meant that naturalists were facing an impossible task in their collecting aims. Natural science quickly fell out of favor in both the amateur collector and the naturalist explorer. Darwin's work, and the emergence of scientific specialties such as botany and ornithology, made natural history collecting an irrelevant exercise, and the reign of the naturalist explorers ended.[341]

At the end of the British Imperial Century exploration for the sake of science was still a major focus. But its demise, or at least major deferral, would come because of something that no one had predicted. As the Imperial Century had begun with the end of a great war, it ended with the outbreak of an even greater one. No longer was the Royal Navy looking for ways to occupy its ships and men; every available resource would be needed in the global conflict that would become known as the Great War. The assistant surgeons would no longer have the luxury of pairing their naval medical duties with those of their government's desire for exploration. From that point forward, medical officers aboard ship would have the wellbeing of the ship's crew as their total responsibility.

By the end of the British Imperial Century the influence of the Scots was universal. Books such as W.J. Rattray's *The Scot in British North America* (1880), pointed to a world dominated by Scottish culture and values. These were held to range from education—that touchstone of the Enlightenment—to the dominance of Scots in the professions, particularly medicine. Academic institutions around the world adopted the intellectual pedagogy that had been founded within the Scottish universities, and at the turn of the 19th/20th century students studying a range of disciplines, from business to law to medicine, were learning in the "Scottish way." [342]

But the very factors that had coalesced to create the medical explorers were no longer unique. The Scottish Enlightenment had created a scientific, literary, and social community that was the envy of Europe. The Enlightenment had enhanced an already strong educational system, providing a model for universal education for boys regardless of their social, economic, or religious status. Developments within the four Scottish universities raised their students above their English and Irish counterparts, and the advancements within the University of Edinburgh, particularly in its Medical School, made it the leading institution in the world. While the medical explorers benefited from the intellectual outcomes of the Enlightenment, so did their equivalents in the other regions. Noting the benefits Scotland gained through its model of universal education and a revamped university system, other countries were soon following suit.

By the beginning of the 20th century universal education for boys had become commonplace throughout the developed world. The innovative system of preparing doctors had proven so successful that programs throughout Europe and America now mimicked those components previously found only in Edinburgh. Scotland had been first, and the rest of the world had taken note. Now the very traits that had made the Edinburgh doctors unique had become the standard for all, and the preparation that the medical explorers had received could be found virtually everywhere throughout Europe and North America. As the need for the medical explorers had been created by factors present only during that 100-year period, when those factors disappeared so did the need for the medical explorers' existence. The story of the medical explorers of Edinburgh is one of the right place, the right time, and the right men. Their almost complete domination of this era could not have happened at any other point in history, and it is likely it will never happen again.

Appendix

Scottish Medical Explorers*

during the British Imperial Century

Name	Birth–Death	University	Area of Exploration
James Edward Aitchison	1835–1898	Edinburgh	India
Thomas Edward Aitchison	1836–1898	Edinburgh	India
John Anderson	1833–1900	Edinburgh	India/Egypt
Thomas Anderson	1832–1870	Edinburgh	India
William Baird	1803–1872	Edinburgh	India/China
William Balfour Baikie	1825–1864	Edinburgh	West Africa
Andrew Balfour	1873–1931	Edinburgh	North Africa
Edward Balfour	1813–1889	Edinburgh	India
Sir Isaac Bayley Balfour	1853–1922	Glasgow	Yemen
Alexander Berry	1781–1873	Edinburgh	Australia
George Bidie	1830–1913	Aberdeen	India
Sir Gilbert Blaine	1749–1834	Glasgow	West Indies
Robert Brown	1773–1858	Edinburgh	Australia/Madeira
Robert Brown	1842–1895	Edinburgh	Canada
David Bruce	1855–1931	Edinburgh	South Africa
William Speirs Bruce	1867–1921	Edinburgh	Arctic/Antarctica
Francis Buchanan-Hamilton	1762–1829	Edinburgh	India
Archibald Campbell	1805–1874	Edinburgh	India
Hugh Cleghorn	1820–1895	Edinburgh	India
Alexander Collie	1793–1835	Edinburgh	China/East Indies
John Crawfurd	1783–1868	Edinburgh	China/East Indies

David Douglas Cunningham	1843–1914	Edinburgh	India
Robert Oliver Cunningham	1841–1918	Edinburgh	South America
Hugh Falconer	1808–1865	Edinburgh	India
Henry Ogg Forbes	1851–1932	Edinburgh	New Zealand
Andrew Thomas Gage	1871–1945	Aberdeen	India
George Gardner	1810–1849	Glasgow	Brazil/Ceylon
Alexander Gibson	1800–1867	Edinburgh	India
John Gillies	1792–1834	Edinburgh	South America
Sir James Hector	1834–1907	Edinburgh	New Zealand
John Imray	1811–1880	Edinburgh	Dominica
William Jack	1795–1822	Edinburgh	Sumatra
William Jameson	1796–1873	Edinburgh	South America
William Jameson	1815–1882	Edinburgh	India
Henry Halcro Johnston	1856–1939	Edinburgh	Sudan/India
George King	1840–1909	Aberdeen	India
John Kirk	1832–1922	Edinburgh	East Africa
William Lockhead	1753–1815	Glasgow	West Indies
David Lyall	1817–1895	Edinburgh	Antarctica
John MacAdam	1827–1865	Edinburgh	Australia
James MacFadyen	1799–1850	Glasgow	Jamaica
Paul MacGillivray	1834–1895	Aberdeen	Australia
Alistair MacKay	1878–1914	Dundee	Arctic/Antarctica
Patrick Mason	1844–1922	Aberdeen	China
Sir James McGrigor	1771–1858	Edinburgh	Spain/Portugal
William McIntosh	1838–1931	Edinburgh	Oceania
Archibald Menzies	1754–1852	Edinburgh	North America
Alexander Morrison	1849–1913	Edinburgh	South Pacific
Neil Gordon Munro	1863–1942	Edinburgh	Japan
Sir John Murray	1841–1914	Edinburgh	Oceania
James Muttlebury	1775–1832	St. Andrews	Jamaica
Walter Oudney	1790–1842	Edinburgh	West Africa
Mungo Park	1771–1806	Edinburgh	West Africa
Sir David Prain	1857–1944	Edinburgh	India
George Hogarth Pringle	1830–1872	Edinburgh	Australia

John Rae	1813–1893	Edinburgh	Canada/Arctic
Sir John Richardson	1787–1865	Edinburgh	Canada/Arctic
William Roxburgh	1751–1815	Edinburgh	India
John Forbes Royle	1798–1858	Edinburgh	India
Helenus Scott	1760–1821	Edinburgh	India
John Scouler	1804–1871	Glasgow	North America
Andrew Sinclair	1796–1861	Glasgow	New Zealand
Andrew Smith	1797–1872	Edinburgh	South Africa
John Lindsay Stewart	1831–1873	Glasgow	India
James Stirton	1833–1917	Edinburgh	Egypt
Peter Cormac Sutherland	1822–1900	St. Andrews	Antarctica
Sir Charles Wyville Thomson	1830–1882	Edinburgh	Oceania
Joseph Thomson	1858–1895	Edinburgh	East Africa
Robert Thomson	1810–1864	Glasgow	India
Thomas Thomson	1817–1878	Glasgow	India
John Forbes Watson	1827–1892	Aberdeen	India
Sir George Watt	1851–1930	Glasgow	India
Thomas Braidwood Wilson	1792–1843	Edinburgh	Australia

* These men were all Scottish. They all studied medicine at one of the Scottish universities, although some took a greater interest in fields such as geology or botany before earning their degrees. Many of them attended more than one Scottish institution. Where that is the case, only the final university they attended is listed, under the assumption that that is where their degree was issued. All were active explorers during the British Imperial Century, and their major exploration efforts were in the areas listed here.

Bibliography

"A History of the Study of Marine Biology" (Retrieved 17 May 2021) Houston, TX: Marine BioConservation Society.

Adams, S. (2014) "Friendship, Loyalty and Allegiance in the Civil War: Scotland, 1637–51," in *Faces of Communities: Social Ties between Trust, Loyalty and Conflict,* S. Feickert, A. Haut and K. Sharaf (eds), Bonn, GR: V & R Press.

Allan, S. and Forsyth, D. (2014) *Common Cause: Commonwealth Scots and the Great War,* Edinburgh, UK: National Museum of Scotland.

Allen, R.B. (2014) *European Slave Trading in the Indian Ocean, 1500–1850,* Athens, OH: Ohio University Press.

Anderson, R.D. (2001) "Universities: 2. 1720–1960", in M. Lynch, ed., *The Oxford Companion to Scottish History,* Oxford, UK: Oxford University Press.

Aronson, T. (1979) *Kings Over the Water: The Saga of the Stuart Pretenders,* London, UK: Cassell Publishers Limited.

Baker, C.A. (1972) "The Development of the Administration to 1897," in *The Early History of Malawi,* edited by Bridglal Pachai, London, UK: Longman.

Barber, L. (1980) *The Heyday of Natural History: 1820–1870,* Garden City, NY: Doubleday and Company, Inc.

Bartholomew, M & Morris, P. (2003) "Science in the Scottish Enlightenment," in *The Rise of Scientific Europe 1500–1800,* eds. David Goodman and Colin A. Russell, London, UK: The Open University.

Berry, C.J. (2013) *The Idea of Commercial Society in the Scottish Enlightenment,* Edinburgh, UK: Edinburgh University Press.

Bierman, J. (1990) *Dark Safari: The Life Behind the Legend of Henry Morton Stanley,* Austin, TX: University of Texas Press.

Blackden, S. (1968) *A Tradition of Excellence: A Brief History of Medicine in Edinburgh,* Edinburgh, UK: Duncan, Flockhart & Co. Ltd.

Bowie, K. (2008) "Popular Resistance, Religion and the Union of 1707," *Scotland and the Union, 1707–2007,* Edinburgh, UK: Edinburgh University Press.

Bowie, K. (2018) "National Opinion and the Press in Scotland Before the Union of 1707," *Scottish Affairs,* 27.1, Edinburgh, UK: Edinburgh University Press.

Brent, P. (1977) *Black Nile: Mungo Park and the Search for the Niger,* London, UK: Gordon & Cremonesi.

Brockway, L.H. (1979) *Science and Colonial Expansion: The Role of the British Royal Botanic Gardens,* London, UK: Academic Press.

Brown, S.R. (2001) *Sightseers and Scholars: Scientific Travelers in the Golden Age of Natural History,* Toronto, CA: Key Porter Books.

Bruce, D.A. (2014) *The Mark of the Scots: Their Astonishing Contributions to History, Science, Democracy, Literature and the Arts,* New York, NY: Skyhorse Publishing.

Bruce, W.S. (1911) *Polar Exploration,* New York, NY: Henry Holt and Company.

Campbell, E. (1999) *Saints and Sea-Kings: The First Kingdom of the Scots,* Edinburgh, UK: Canongate Books, Ltd.

Campbell, R.H. (1964) "The Anglo-Scottish Union of 1707. II: The Economic Consequences", *Economic History Review,* vol. 16, April.

Canada Drainage Basins (1985) *The National Atlas of Canada,* 5th edition. Natural Resources Canada.

Cannadine, D. (2017) *Victorious Century: The United Kingdom, 1800–1906,* New York, NY: Penguin Books.

Cherry, S, (1966) *Medical Services and the Hospitals in Britain, 1860–1939,* Cambridge, UK: Cambridge University Press.

Chitnis, A.C. (1976) *The Scottish Enlightenment,* London, UK: Croom Helm Ltd.

Colley, L. (1992) Britons Forging the Nation 1707–1873, New Haven, CN: Yale University Press.

Cooper, G. (1922) "The Last Lord Camelford," *The Mariner's Mirror,* 8, (6).

Corfield, R, (2003) *The Silent Landscape: The Scientific Voyage of HMS Challenger,* Washington, DC: Joseph Henry Press.

Coupland, R. (Sir) (1967) *The Exploitation of East Africa 1856–1890: The Slave Trade and the Scramble,* Evanston, IL: Northwestern University Press.

Davidson, B. (1964) *The African Past: Chronicles from Antiquity to Modern Times,* London, UK: Longmans, Green and Co, Ltd.

Davis, C. and Lamb, W.K. (1997) *Greater Vancouver Book: An Urban Encyclopedia,* Surrey, BC: Linkman Press.

Desmond, R. (2007) *The History of the Royal Botanic Gardens Kew,* London, UK: Kew Publishing.

Dingwall, H.M. (2003) *A History of Scottish Medicine: Themes and Influences,* Edinburgh, UK: Edinburgh University Press.

Endersby, J. (2019) "Gardens of the Empire: Kew and the Colonies," Gresham College Presentation, 2 December.

Faguy, A. (2024) DNA test helps identify sailor from doomed Arctic expedition, BBC.com/news/article, 24 September 2024.

Fry, P.S. (1990) *The Kings and Queens of England and Scotland,* London, UK: Dorling Kindersley Limited.

Fry, P.S. and Fry, F. (1985) *The History of Scotland,* London, UK: Ark Paperbacks.

Livingston, M. (2011) *The Battle of Brunanburh: A Casebook,* Exeter, UK: University of Exeter Press.

Gordon, R. M (2009) *The Infamous Burke and Hare,* London, UK: McFarland & Company, Inc.

Goring, R. (2008) *Scotland: The Autobiography,* New York, NY: The Overlook Press.

Grace, R.J. (2014) *Opium and Empire: The Lives and Careers of William Jardine and James Matheson,* Montreal, CA: McGill-Queens University Press.

Grant, J. (ed.) (1914) "The Old Scots Navy from 1689 to 1710," Publications of the Navy Records Society 44, London, UK: Navy Records Society.

Herman, A. (2001) *How the Scots Invented the Modern World,* New York, NY: Broadway Books.

Hibbert, C. (1982) *Africa Explored: Europeans in the Dark Continent, 1769–1889,* London, UK: W.W. Norton & Company.

Howard, C. (ed.) (1951) *West African Explorers.* London, UK: Oxford University Press.

Israel, J. (2011) *Democratic Enlightenment: Philosophy, Revolution, and Human Rights 1750–1790,* London, UK: Oxford University Press.

Jeal, T. (1973) *Livingstone,* New York, NY: G.P. Putnam's Sons.

Jeal, T. (2013) *Livingstone: Revised and Expanded Edition,* New Haven, CT: Yale University Press.

Jenkinson, J. (1993) *Scottish Medical Societies: 1731–1939 Their History and Records,* Edinburgh, UK: Edinburgh University Press.

Kamm, A. (2013) *Scottish Explorers,* Edinburgh, UK: National Museums of Scotland.

Kaufmann, C.D. & Pape, R.A. (1999) "Explaining Costly International Moral Action: Britain's Sixty-Year Campaign Against the Atlantic Slave Trade," *International Organization,* Boston, MA: MIT Press.

Keevil, J.J. (1948) "Archibald Menzies 1754–1852," *Bulletin of the History of Medicine,* Vol. 22, No. 6, Baltimore, MD: Johns Hopkins University Press.

Bibliography

Kryza, F.T. (2006) *The Race for Timbuktu: In Search of Africa's City of Gold,* New York: Harper Collins.

Lagemann, E.C. (1992) *The Politics of Knowledge: The Carnegie Corporation, Philanthropy and Public Policy,* Chicago, IL: University of Chicago Press.

Leyburn, J.G. (1962) *The Scotch-Irish: A Social History,* Chapel Hill, NC: University of North Carolina Press.

Liebowitz, D. (1999) *The Physician and the Slave Trade: John Kirk, the Livingstone Expeditions, and the Crusade Against Slavery in East Africa,* New York, NY: W.H. Freeman and Company.

Lindsay, A. (2008) *Seeds of Blood and Beauty: Scottish Plant Explorers,* Edinburgh, UK: Birlinn Limited.

Linklater, E. (1972) *The Voyage of the Challenger,* London, UK: John Murry Publishers, Ltd.

Lysaght, A.M. (1971) *Joseph Banks in Newfoundland and Labrador, 1766,* Berkeley, CA: University of California Press.

MacDougall, D. (2019) End*less Novelties of Extraordinary Interest: The Voyage of H.M.S. Challenger and the Birth of Modern Oceanography,* New Haven, CT: Yale University Press.

MacLean, F. (Sir) (2019) *Scotland: A Concise History,* London, UK: Thames & Hudson, Ltd.

McCabe, M. (2013) "On the 1707 Union," Work from Private Study.

McCarthy, J. (2008) *Monkey Puzzle Man: Archibald Menzies, Plant Hunter,* Dunbeath, UK: Whittles Publishing.

McEwen, R, (2012) "The Remarkable Botanist Physicians: Natural Science in the Age of Empire," in www.electricscotland.com.

McGoogan, K. (2002) *Fatal Passage: The Story of John Rae, the Arctic Hero Time Forgot,* New York, NY: Carroll & Graff Publishers.

McLeod, J. (2007) *The Routledge Companion to Postcolonial Studies,* New York, NY: Routledge Publishing.

Markham, C.R. (1881) *The Fifty Years' Work of the Royal Geographical Society,* London, UK: John Murray.

Maxwell, G. (1998) *A Gathering of Eagles: Scenes from Roman Scotland,* Edinburgh, UK: Canongate Books, Ltd.

Maxwell, I. (2019) *Tracing Your Scottish Ancestors: A Guide for Family Historians,* Havertown, PA: Pen and Sword.

Menzies to Banks, 21 August 1786. Banks correspondence, Royal Botanic Gardens, Kew.

Menzies to Banks, 7 February 1795. Banks correspondence, Royal Botanic Gardens, Kew.

Merrill, L.L. (1989) *The Romance of Victorian Natural History,* London, UK: Oxford University Press.

Moorehead, A. (1960) *The White Nile,* London, UK: Harper & Row Publishers.

Murray, J. (2018) *Report on the Scientific Results of the Voyage of H.M.S. Challenger during the Years 1873–76 under the Command of Captain George S. Nares and the Late Captain Frank Tourle Thomson,* New York, NY: Franklin Classics.

Newman, P.C. (1985) *Company of Adventurers,* Vol. l, Markham, Ontario, CA: Viking Penguin Books of Canada.

Nicol, D. (1956) "Mungo Park and the River Niger," *African Affairs* 55, no. 218.

Niezgoda, C. (2011) "Focus: Economic Botany," Field Museum Presentation, 10 January.

Olson, W. M (1993) *The Alaska Travel Journal of Archibald Menzies 1793–1794,* Fairbanks, AK: University of Alaska Press.

Parsons, T.H. (2019) *British Imperial Century, 1815–1914: A World History Perspective,* Lanham, MD: Rowman & Littlefield Publishers.

Phillipson, N. (1973) "Towards a definition of the Scottish Enlightenment," in P. Fritz and D. Williams (eds), *City and Society in the Eighteenth Century,* Toronto, CA: Hakkert.

O'Brian, P. (1977) *Joseph Banks: A Life.* Chicago, IL: University of Chicago Press.

Qman, C. (Sir) (1923) *England in the Nineteenth Century,* New York, NY: Longmans, Green and Co.

Pearce, F.B. (1920) *Zanzibar: The Island Metropolis of Eastern Africa,* New York, NY: Dutton and Company.

Pettitt, C. (2013) *Dr Livingstone I Presume: Missionaries, Journalists, Explorers and Empire,* London, UK: Profile Books.

Rae, J. (2012) *The Arctic Journals of John Rae,* Vancouver, CN: Touchwood Editions.

Raffensperger, J. (2011) *A Brief History of the Edinburgh School of Medicine,* New York, NY: Cosimo Classics.

Ransford, O. (1978) *David Livingstone: The Dark Interior,* London, UK: John Murray.

Rosner, L. (1991) *Medical Education in the Age of Improvement,* Edinburgh, UK: Edinburgh University Press.

Ross, A. (1974) *Pagan Celtic Britain: Studies in Iconography and Tradition,* London, UK: Sphere Books Ltd.

Rousseau, J.J. (1979) *Allan Bloom Translator, Emile: Or, On Education,* New York, NY: Basic Books.

"Royal Geographical Society – History," (2014) Royal Geographical Society.

Royle, T. (2004) *Civil War: The War of the Three Kingdoms 1638–1660,* New York, NY: Little, Brown Book Group.

Scott, W. (Sir) (1969) *The Tales of a Grandfather* (abr.), Elizabeth W. Grierson, London, UK: A. &. C. Black Ltd.

Shillington, K. (2005) *History of Africa,* New York: Macmillan Publishers Limited.

Smith, N. (2022) Summary, Analysis & Insights on Emile by Rousseau, Articlemyriad.com.

Speak, P. (2003) *William Speirs Bruce: Polar Explorer and Scottish Nationalist,* Edinburgh, UK: NMS Publishing.

Special Collections, University of Edinburgh, (1973) Attendance General 1647/46/5.

Stevenson, D. (1973) *The Scottish Revolution 1637–1644,* Newton Abbot, UK: David and Charles.

Tanner, R. (2009) "Franco-Scottish Alliance," *The Oxford Companion to British History,* London, UK: University Press.

Thomson, C.W. (2014) *The Depths of the Sea: An Account of the General Results of the Dredging Cruises of H.M. SS. 'Porcupine' and 'Lightning' during the Summers of 1868, 1869, and 1870, under the Scientific Direction of Dr. Carpenter, J. Gwyn Jeffreys, and Dr. Wyville Thomson,* Charleston, SC: Nabu Press.

Turnbull, F. (1954) "Vancouver and Menzies or Medicine on the Quarterdeck," *Bulletin of the Vancouver Medical Association,* April, Vancouver, CA.

Whelan, K. (2022) *The Other Within: Ireland, Britain and the Act of Union,* South Bend, IN: University of Notre Dame.

Wild, H. (1960–61) "Sir John Kirk," *Kirkia: Zimbabwean Journal of Botany,* Vol. 1, Harare, ZW: Government Printers.

Williams, I. P and Dudeney, J. (2018) *William Speirs Bruce: Forgotten Polar Hero,* Gloucestershire, UK: Amberley Publishing.

Wilson, M. (2013) "Riddle of how the Monkey Puzzle Tree came to be a UK Favourite," *Financial Times,* London, UK, 5 July.

Womersley, T. & Crawford, D.H. (2010) *Bodysnatchers to Lifesavers: Three Centuries of Medicine in Edinburgh,* Edinburgh, UK: Luath Press Ltd.

Woodruff, D. (1939) *The Story of the British Colonial Empire,* London, UK: William Clowes & Sons Ltd.

Wright, E. (2008) Lost *Explorers: Adventurers who Disappeared Off the Face of the Earth,* Crow's Nest, AU: Allen & Unwin.

Zappala, C. (2013) "Conquest, Absorption or Something Else?" Swansea, UK: Trinity Saint David, Essay.

Endnotes

Introduction

1   Sir Charles Qman, *England in the Nineteenth Century*, (New York, NY: Longmans, Green and Co., 1923), p. 53.

2   David Cannadine, *Victorious Century: The United Kingdom*, 1800–1906, (New York, NY: Penguin Books, 2017), pp. 100–109.

3   Qman, *England in the Nineteenth Century*, p. 56.

4   Kevin Whelan, *The Other Within: Ireland, Britain and the Act of Union*, (South Bend, IN: University of Notre Dame, 2022), p. 1.

5   Duncan A. Bruce, *The Mark of the Scots: Their Astonishing Contributions to History, Science Democracy, Literature and the Arts*, (New York, NY: Skyhorse Publishing, 2014), p. 59.

6   Lisa Rosner, *Medical Education in the Age of Improvement*, (Edinburgh, UK: Edinburgh University Press, 1991), p. 11.

Chapter 1

7   Michael Livingston, *The Battle of Brunanburh: A Casebook*, (Exeter, UK: University of Exeter Press, 2011), p. 1.

8   Sir Walter Scott, *The Tales of a Grandfather*, (abr.) Elizabeth W. Grierson, (London, UK: A. &. C. Black, Ltd., 1969), pp. 47–53.

9   Helen M. Dingwall, *A History of Scottish Medicine: Themes and Influences*, (Edinburgh, UK: Edinburgh University Press, 2003), pp. 51–55.

10 Ibid, p. 58 and p. 72.

11 Roland Tanner, "Franco-Scottish Alliance," *The Oxford Companion to British History*, (London, UK: University Press, 2009), pp. 390–91.

12 Dingwall, *A History of Scottish Medicine: Themes and Influences*, pp. 10–21.

13 Plantagenet Somerset Fry, *The Kings and Queens of England and Scotland*, (London, UK: Dorling Kindersley Limited, 1990), pp. 40–55.

14 Mark McCabe, *On the 1707 Union*, (Work from Private Study, 2013).

15 Sir Fitzroy MacLean, *Scotland: A Concise History*, (London, UK: Thames & Hudson, Ltd. 2019), p. 71.

16 David Stevenson, *The Scottish Revolution 1637–1644*, (Newton Abbot, UK: David and Charles, 1973).

17 MacLean, *Scotland: A Concise History*, pp. 82–84.

18 Sharon Adams, "Friendship, Loyalty and Allegiance in the Civil War: Scotland, 1637–51," in *Faces of Communities: Social Ties between Trust, Loyalty and Conflict*, S. Feickert, A. Haut and K. Sharaf (eds), (Bonn, GR: V & R Press, 2014), pp. 9–24.

19 Trevor Royle, *Civil War: The War of the Three Kingdoms 1638–1660*, (New York, NY: Little, Brown Book Group, 2004).

20 Theo Aronson, *Kings Over the Water: The Saga of the Stuart Pretenders*, (London, UK: Cassell Publishers Limited, 1979), pp. 7–16.

21 MacLean, *Scotland: A Concise History*, p. 90.

22 Scott, *The Tales of a Grandfather*, pp. 218–220.

23 Aronson, *Kings Over the Water: The Saga of the Stuart Pretenders*, pp. 29–39.

24 McCabe, *On the 1707 Union.*

25 Scott, *The Tales of a Grandfather*, pp. 239–241.

26 McCabe, *On the 1707 Union.*

27 Karin Bowie, "Popular Resistance, Religion and the Union of 1707," *Scotland and the Union, 1707–2007*, (Edinburgh, UK: Edinburgh University Press, 2008).

28 Linda Colley, *Britons Forging the Nation 1707–1873*, (New Haven, CN: Yale University Press, 1992).

29 R. H. Campbell, "The Anglo-Scottish Union of 1707. II: The Economic Consequences", *Economic History Review*, vol. 16, April 1964.

30 Dingwall, *A History of Scottish Medicine: Themes and Influences*, pp. 21–22.

31 Tara Womersley and Dorothy H. Crawford, *Bodysnatchers to Lifesavers: Three Centuries of Medicine in Edinburgh*, (Edinburgh, UK: Luath Press Ltd., 2010), pp. 20–21.

32 Dingwall, *A History of Scottish Medicine: Themes and Influences*, pp. 68–69.

33 Christopher J. Berry, *The Idea of Commercial Society in the Scottish Enlightenment*, (Edinburgh, UK: Edinburgh University Press, 2013), p. 2.

34 Rosemary Goring, *Scotland: The Autobiography*, (New York, NY: The Overlook Press, 2008), p. 165.

35 Peter and Fiona Somerset Fry, *The History of Scotland*, (London, UK: Ark Paperbacks, 1985), p. 82.

36 Campbell, "The Anglo-Scottish Union of 1707. II: The Economic Consequences", *Economic History Review.*

37 Berry, *The Idea of Commercial Society in the Scottish Enlightenment*, pp. 2–3.

38 Carla Zappala, *Conquest, Absorption or Something Else?* (Swansea, UK: Trinity Saint David, Essay, 2013).

39 Karin Bowie, "National Opinion and the Press in Scotland Before the Union of 1707," *Scottish Affairs*, 27.1, (Edinburgh, UK: Edinburgh University Press, 2018), pp. 13–19.

Chapter 2

40 Jean Jacques Rousseau, Allan Bloom Translator, *Emile: Or, On Education*, (New York, NY: Basic Books, 1979).

41 Nicole Smith, Summary, Analysis & Insights on Emile by Rousseau, (Articlemyriad.com, 2022).

42 Arthur Herman, *How the Scots Invented the Modern World*, (New York, NY: Broadway Books, 2001), pp. 63–64.

43 Somerset Fry and Somerset Fry, *The History of Scotland*, pp. 205–206.

44 Berry, *The Idea of Commercial Society in the Scottish Enlightenment*, p. 1.

45 Anand C. Chitnis, *The Scottish Enlightenment*, (London, UK: Croom Helm Ltd. 1976), pp. 4–11.

46 Nicholas Phillipson, "Towards a definition of the Scottish Enlightenment," in p. Fritz and D. Williams (eds), *City and*

Society in the Eighteenth Century, (Toronto, Hakkert, 1973).

47   Dingwall, *A History of Scottish Medicine: Themes and Influences*, p. 5.

48   Ibid., p. 156.

49   Ian Maxwell, *Tracing Your Scottish Ancestors: A Guide for Family Historians*, (Haverstown, PA: Pen and Sword, 2019), pp. 38–44.

50   Stuart Allan and David Forsyth, *Common Cause: Commonwealth Scots and the Great War*, (Edinburgh, UK: National Museum of Scotland, 2014), p. 7.

51   Chitnis, *The Scottish Enlightenment*, pp. 5–7.

52   Anne Ross, *Pagan Celtic Britain: Studies in Iconography and Tradition*, (London, UK: Sphere Books Ltd. 1974).

53   Gordon Maxwell, *A Gathering of Eagles: Scenes from Roman Scotland*, (Edinburgh, UK: Canongate Books, Ltd., 1998) p. 56.

54   Dingwall, *A History of Scottish Medicine: Themes and Influences*, pp. 64–66.

55   Ewen Campbell, *Saints and Sea-Kings: The First Kingdom of the Scots*, (Edinburgh, UK: Canongate Books, Ltd., 1999), p. 11.

56   Berry, *The Idea of Commercial Society in The Scottish Enlightenment*, p. 5.

57   Somerset Fry and Somerset Fry, *The History of Scotland*, pp. 182–183.

58   Goring, *Scotland: The Autobiography*, p. 228.

59   Chitnis, *The Scottish Enlightenment*, pp. 43–67.

60   Somerset Fry and Somerset Fry, *The History of Scotland*, p. 215.

61   Berry, *The Idea of Commercial Society in the Scottish Enlightenment*, p. 4.

62   Richard J. Grace, *Opium and Empire: The Lives and Careers of William Jardine and James Matheson*, (Montreal, CA: McGill-Queens University Press, 2014), p. 9.

63   Chitnis, *The Scottish Enlightenment*, pp. 75–77.

64   Ibid., pp. 78–86.

65   Somerset Fry and Somerset Fry, *The History of Scotland*, p. 216.

66   Maxwell, Tracing Your Scottish Ancestors: A Guide for Family Historians, p. 76.

67   Ibid., p. 77.

68   Ibid., pp. 70–73.

69   Chitnis, *The Scottish Enlightenment*, p. 4.

70   Maxwell, *Tracing Your Scottish Ancestors: A Guide for Family Historians*, p. 79.

71   Dingwall, *A History of Scottish Medicine: Themes and Influences*, pp. 69–70.

72   Herman, *How Scots Invented the Modern World*, p. 25.

73   Ibid., pp. 25–26.

74   R. D. Anderson, "Universities: 2. 1720–1960", in M. Lynch, ed., *The Oxford Companion to Scottish History* (Oxford, UK: Oxford University Press, 2001), pp. 612–14.

75   Michael Bartholomew and Peter Morris, "Science in the Scottish Enlightenment," in *The Rise of Scientific Europe 1500–1800*, eds. David Goodman and Colin A. Russell, (London, UK: The Open University, 2003).

76   Ellen Condliffe Lagemann, *The Politics of Knowledge: The Carnegie Corporation, Philanthropy and Public Policy*, (Chicago, IL: University of Chicago Press, 1992) p. 17.

77   Chitnis, *The Scottish Enlightenment*, pp. 16–33.

78 Jonathan Israel, *Democratic Enlightenment: Philosophy, Revolution, and Human Rights 1750–1790*, (London, UK: Oxford University Press, 2011). p. 233.

79 Bartholomew and Morris, "Science in the Scottish Enlightenment," in *The Rise of Scientific Europe 1500–1800*.

80 Chitnis, *The Scottish Enlightenment*, pp. 195–197.

81 Grace, *Opium and Empire: The Lives and Careers of William Jardine and James Matheson*, p. 29.

82 R. Michael Gordon, *The Infamous Burke and Hare*, (London, UK: McFarland & Company, Inc. 2009), p. 5.

83 James G. Leyburn, *The Scotch-Irish: A Social History*, (Chapel Hill, NC: University of North Carolina Press, 1962), p. 45.

Chapter 3

84 Dingwall, *A History of Scottish Medicine: Themes and Influences*, pp. 3–4.

85 Ibid., p. 43.

86 Womersley and Crawford, *Bodysnatchers to Lifesavers: Three Centuries of Medicine in Edinburgh*, p. 24.

87 Dingwall, *A History of Scottish Medicine: Themes and Influences*, p. 45.

88 Stephanie Blackden, *A Tradition of Excellence: A Brief History of Medicine in Edinburgh*, (Edinburgh, UK: Duncan, Flockhart & Co. Ltd., 1968), p. 5.

89 Womersley and Crawford, *Bodysnatchers to Lifesavers: Three Centuries of Medicine in Edinburgh*, p. 21.

90 Dingwall, *A History of Scottish Medicine: Themes and Influences*, p. 76.

91 Womersley and Crawford, *Bodysnatchers to Lifesavers: Three Centuries of Medicine in Edinburgh*, pp. 24–25.

92 Rosner, *Medical Education in the Age of Improvement*, pp. 88–89.

93 Grace, *Opium and Empire: The Lives and Careers of William Jardine and James Matheson*, p. 18.

94 Blackden, *A Tradition of Excellence: A Brief History of Medicine in Edinburgh*, p. 5.

95 Rosner, *Medical Education in the Age of Improvement*.

96 Dingwall, *A History of Scottish Medicine: Themes and Influences*, p. 48.

97 Rosner, *Medical Education in the Age of Improvement*, pp. 86–90.

98 Ibid.

99 John Raffensperger, *A Brief History of the Edinburgh School of Medicine*, (New York, NY: Cosimo Classics, 2011), p. 75.

100 Blackden, *A Tradition of Excellence: A Brief History of Medicine in Edinburgh*, p. 4.

101 Gordon, *The Infamous Burke and Hare*, pp. 7–46.

102 Womersley and Crawford, *Bodysnatchers to Lifesavers: Three Centuries of Medicine in Edinburgh*, pp. 32–38.

103 Blackden, *A Tradition of Excellence: A Brief History of Medicine in Edinburgh*, p. 5.

104 Womersley and Crawford, *Bodysnatchers to Lifesavers: Three Centuries of Medicine in Edinburgh*, pp. 25–26.

105 Blackden, *A Tradition of Excellence: A Brief History of Medicine in Edinburgh*, p. 5.

106 Ibid., p. 9.

107 Steven Cherry, *Medical Services and the hospitals in Britain, 1860–1939*, (Cambridge, UK: Cambridge University Press, 1966), pp. 1–2.

108 Womersley and Crawford, *Bodysnatchers to Lifesavers: Three Centuries of Medicine in Edinburgh*, p. 25.

109 Ibid., p. 26.

Endnotes

110 Jacqueline Jenkinson, *Scottish Medical Societies: 1731–1939 Their History and Records*, (Edinburgh, UK: Edinburgh University Press, 1993), pp. 4–20.

111 Ibid., pp. 4–14.

112 Rosner, *Medical Education in the Age of Improvement*, p. 93.

113 Cherry, *Medical Services and the Hospitals in Britain, 1860–1939*, p. 27.

114 Grace, *Opium and Empire: The Lives and Careers of William Jardine and James Matheson*, p. 17.

115 Blackden, *A Tradition of Excellence: A Brief History of Medicine in Edinburgh*, p. 5.

116 Womersley and Crawford, *Bodysnatchers to Lifesavers: Three Centuries of Medicine in Edinburgh*, pp. 22–23.

117 Dingwall, *A History of Scottish Medicine: Themes and Influences*, p. 83.

118 Cherry, *Medical Services and the Hospitals in Britain, 1860–1939*, pp. 17–18.

119 Gordon, *The Infamous Burke and Hare*, p. 1.

120 Raffensperger, *A Brief History of the Edinburgh School of Medicine*, p. 76.

121 Bruce, *The Mark of the Scots: Their Astonishing Contributions to History, Science Democracy, Literature and the Arts*, p. 25.

122 Raffensperger, *A Brief History of the Edinburgh School of Medicine*, p. 75.

Chapter 4

123 Campbell, *Saints and Sea-Kings: The First Kingdom of the Scots*, p. 9.

124 Leyburn, *The Scotch-Irish: A Social History*, p. 45.

125 Allan and Forsyth, *Common Cause: Commonwealth Scots and the Great War*, p. 2.

126 Peter and Fiona Somerset Fry, *The History of Scotland*, p. 219.

127 Allan and Forsyth, *Common Cause: Commonwealth Scots and the Great War*, p. 14.

128 J. Grant (ed.), "The old Scots navy from 1689 to 1710," *Publications of the Navy Records Society* 44, (London, UK: Navy Records Society, 1914), p. 448.

129 Grace, *Opium and Empire: The Lives and Careers of William Jardine and James Matheson*, p. 9.

130 Maxwell, *Tracing Your Scottish Ancestors: A Guide for Family Historians*, p. 122.

131 Womersley and Crawford, *Bodysnatchers to Lifesavers: Three Centuries of Medicine in Edinburgh*, p. 27.

132 Whelan, *The Other Within: Ireland, Britain and the Act of Union*, p. 1.

133 Ann Lindsay, *Seeds of Blood and Beauty: Scottish Plant Explorers*, (Edinburgh, UK: Birlinn Limited, 2008), p. 113.

134 Ron McEwen, "The Remarkable Botanist Physicians: Natural Science in the Age of Empire," (in www.electricscotland.com., 2012).

135 Eric Linklater, *The Voyage of the Challenger*, (London, UK: John Murry Publishers, Ltd., 1972), p. 12.

136 Timothy H. Parsons, *British Imperial Century, 1815–1914: A World History Perspective*, (Lanham, MD: Rowman & Littlefield Publishers, 2019), pp. 2–4.

137 Douglas Woodruff, *The Story of the British Colonial Empire*, (London, UK: William Clowes & Sons Ltd. 1939), pp. 19–20.

138 Timothy H. Parsons, *The British Imperial Century, 1815–1914: A World History Perspective*, (Lanham, MD: Rowman & Littlefield Publishers, 2019), pp. 14–15.

139 Bruce, *The Mark of the Scots: Their Astonishing Contributions to History, Science, Democracy, Literature, and the Arts*, pp. 83–84.

140 Grace, *Opium and Empire: The Lives and Careers of William Jardine and James Matheson*, pp. 9–11.

141 Timothy H. Parsons, *The British Imperial Century, 1815–1914: A World History Perspective*, (Lanham, MD: Rowman & Littlefield Publishers, 2019), p. 4.

142 "Canada Drainage Basins," *The National Atlas of Canada*, 5th edition. (Natural Resources Canada. 1985). Retrieved 24 November 2010.

143 Peter C. Newman, *Company of Adventurers*, Vol. l, (Markham, Ontario, CA: Viking Penguin Books of Canada, 1985).

144 Bruce, *The Mark of the Scots: Their Astonishing Contributions to History, Science, Democracy, Literature, and the Arts*, p. 72.

145 Ibid., p. 71.

146 Ibid., pp. 72–73.

Chapter 5

147 Stephen R. Brown, *Sightseers and Scholars: Scientific Travelers in the Golden Age of Natural History*, (Toronto, CA: Key Porter Books, 2002), p. 72.

148 Averil M. Lysaght, *Joseph Banks in Newfoundland and Labrador, 1766*, (Berkeley, CA: University of California Press, 1971) p. 168.

149 Desmond, Kew: *The History of the Royal Botanic Garden*, pp. 86-87.

150 Patrick O'Brian, *Joseph Banks: A Life*. (Chicago, IL: University of Chicago Press, 1997).

151 Ray Desmond, *The History of the Royal Botanic Gardens Kew*, (London, UK: Kew Publishing, 2007).

152 Lynn L. Merrill, *The Romance of Victorian Natural History*, (London, UK: Oxford University Press, 1989), pp. 8–11.

153 Lynn Barber, *The Heyday of Natural History: 1820–1870*, (Garden City, NY: Doubleday and Company, Inc., 1980), pp. 13–14.

154 Brown, *Sightseers and Scholars: Scientific Travelers in the Golden Age of Natural History*, pp. 12–13.

155 Barber, *The Heyday of Natural History: 1820–1870*, p. 30.

156 Merrill, *The Romance of Victorian Natural History*, p. 12.

157 Womersley and Crawford, *Bodysnatchers to Lifesavers: Three Centuries of Medicine in Edinburgh*, p. 109.

158 Lindsay, *Seeds of Blood and Beauty: Scottish Plant Explorers*, p. 1.

159 Desmond, *The History of the Royal Botanic Gardens Kew*, p. 73.

160 Christine Niezgoda, Focus: Economic Botany, (Field Museum Presentation, 10 January 2011.

161 McEwen, "The Remarkable Botanist Physicians: Natural Science in the Age of Empire."

162 Lindsay, *Seeds of Blood and Beauty: Scottish Plant Explorers*, p. 146.

163 Desmond, *The History of the Royal Botanic Gardens Kew*.

164 Ibid., p. 123.

165 Ibid., pp. 124–125.

166 Jim Endersby, *Gardens of the Empire: Kew and the Colonies*, (Gresham College Presentation, 2 December 2019).

167 Desmond, *The History of the Royal Botanic Gardens Kew*, p. 126.

168 Lucile H. Brockway, *Science and Colonial Expansion: The Role of the British Royal Botanic Gardens*, (London, UK: Academic Press, 1979).

169 Desmond, *The History of the Royal Botanic Gardens Kew*, pp. xiii–xv.

170 McEwen, "The Remarkable Botanist Physicians: Natural Science in the Age of Empire."

171 Endersby, *Gardens of the Empire: Kew and the Colonies*.

172 Clements R. Markham, *The Fifty Years' Work of the Royal Geographical Society*, (London, UK: John Murray, 1881) pp. 7–10.

173 Frank T. Kryza, *The Race for Timbuktu: In Search of Africa's City of Gold*, (New York: HarperCollins, 2006), p. 11.

174 Basil Davidson, *The African Past: Chronicles from Antiquity to Modern Times*, (London, UK: Longmans, Green and Co, Ltd, 1964), pp. 271–272.

175 Christopher Hibbert, Africa *Explored: Europeans in the Dark Continent, 1769–1889*, (London, UK: W. W. Norton & Company, 1982), p. 18.

176 Kryza, *The Race for Timbuktu: In Search of Africa's City of Gold*, p. 11.

177 Peter Brent, *Black Nile: Mungo Park and the Search for the Niger*, (London, UK: Gordon & Cremonesi, 1977), p. 44.

178 Davidson Nicol, "Mungo Park and the River Niger," *African Affairs* 55, no. 218, January 1956, p. 47.

179 Howard (ed.), *West African Explorers*, p. 9.

180 Kryza, *The Race for Timbuktu: In Search of Africa's City of Gold*, pp. 15–18.

181 Ibid., p. 20.

182 "Royal Geographical Society—History," Royal Geographical Society. Retrieved 9 December 2014.

183 Markham, *The Fifty Years' Work of the Royal Geographical Society*, p. 2.

Chapter 6

184 James McCarthy, *Monkey Puzzle Man: Archibald Menzies Plant Hunter*, (Dunbeath, UK: Whittles Publishing, 2008), pp. 4–5.

185 Lindsay, *Seeds of Blood and Beauty: Scottish Plant Explorers*, pp. 5–6.

186 John Joyce Keevil, "Archibald Menzies 1754–1852," *Bulletin of the History of Medicine*, Vol. 22, No. 6, (Baltimore, MD: Johns Hopkins University Press, 1948), p. 796.

187 Lindsay, *Seeds of Blood and Beauty: Scottish Plant Explorers*, pp. 109–111.

188 Ibid., p. 113.

189 McCarthy, *Monkey Puzzle Man: Archibald Menzies Plant Hunter*, pp. 35–38.

190 Ibid., p. 38.

191 Keevil, "Archibald Menzies 1754–1852," *Bulletin of the History of Medicine*, Vol. 22, No. 6, pp 797–798.

192 Lindsay, *Seeds of Blood and Beauty: Scottish Plant Explorers*, pp. 114–115.

193 Menzies to Banks, 21 August 1786. Banks correspondence, Royal Botanic Gardens Kew.

194 McCarthy, *Monkey Puzzle Man: Archibald Menzies Plant Hunter*, p. 54.

195 Menzies to Banks, 7 September 1786. Banks correspondence, Royal Botanic Gardens Kew.

196 McCarthy, *Monkey Puzzle Man: Archibald Menzies Plant Hunter*, p. 54.

197 Ibid., pp. 55–58.

198 Ibid. pp. 64–68.

199 Ibid., p. 71.

200 Chuck Davis & W. Kaye Lamb, *Greater Vancouver Book: An Urban Encyclopedia*, (Surrey, BC: Linkman Press, 1997), pp. 34–36.

201 McCarthy, *Monkey Puzzle Man: Archibald Menzies Plant Hunter*, pp. 75–76.

202 Ibid., p. 72.

203 William M. Olson, *The Alaska Travel Journal of Archibald Menzies 1793–1794*, (Fairbanks, AK: University of Alaska Press, 1993), p. 9.

204 Ibid., p. 10.

205 Keevil, "Archibald Menzies 1754–1852," *Bulletin of the History of Medicine*, Vol. 22, No. 6, p. 799.

206 McCarthy, *Monkey Puzzle Man: Archibald Menzies Plant Hunter*, pp. 72–77.

207 Olson, *The Alaska Travel Journal of Archibald Menzies 1793–1794*, pp. 9–10.

208 Keevil, "Archibald Menzies 1754–1852," *Bulletin of the History of Medicine*, Vol. 22, No. 6, pp. 799–800.

209 McCarthy, *Monkey Puzzle Man: Archibald Menzies Plant Hunter*, p. 82.

210 Ibid., pp. 92–93.

211 Ibid., p. 100.

212 Ibid., pp. 140–142.

213 Keevil, "Archibald Menzies 1754–1852," *Bulletin of the History of Medicine*, Vol. 22, No. 6, pp. 799–800.

214 McCarthy, *Monkey Puzzle Man: Archibald Menzies Plant Hunter*, pp. 109–113.

215 Frank Turnbull, "Vancouver and Menzies or Medicine on the Quarterdeck," *Bulletin of the Vancouver Medical Association*, April , (Vancouver, CA, 1954), pp. 277–285.

216 Menzies to Banks, 7 February 1795. Banks correspondence, Royal Botanic Gardens Kew.

217 Keevil, "Archibald Menzies 1754–1852," *Bulletin of the History of Medicine*, Vol. 22, No. 6, pp. 804–805.

218 Matthew Wilson, "Riddle of how the Monkey Puzzle Tree Came to be a UK Favourite," *Financial Times*, (London, UK, 5 July 2013).

219 Olson, *The Alaska Travel Journal of Archibald Menzies 1793–1794*, p. 19.

220 Keevil, "Archibald Menzies 1754–1852," *Bulletin of the History of Medicine*, Vol. 22, No. 6, pp. 801–804.

221 Guy Cooper, "The Last Lord Camelford," *The Mariner's Mirror*, 8, (6), 1922.

222 McCarthy, *Monkey Puzzle Man: Archibald Menzies Plant Hunter*, p. 188.

223 Olson, *The Alaska Travel Journal of Archibald Menzies 1793–1794*, pp. 21–22.

224 Ibid., p. 5.

225 Keevil, "Archibald Menzies 1754–1852," *Bulletin of the History of Medicine*, Vol. 22, No. 6, p. 801.

226 Ibid., p. 809.

Chapter 7

227 Chaim D. Kaufmann and Robert A. Pape, "Explaining Costly International Moral Action: Britain's Sixty-Year Campaign Against the Atlantic Slave Trade," *International Organization*, (Boston, MA: MIT Press, 1999), pp. 631–668.

228 Cannadine, *Victorious Century: The United Kingdom, 1800–1906*, p. 187.

229 Daniel Liebowitz, *The Physician and the Slave Trade: John Kirk, the Livingstone Expeditions, and the Crusade Against Slavery in East Africa*, (New York, NY: W. H. Freeman and Company, 1999), p. 10.

230 H. Wild, "Sir John Kirk," *Kirkia: Zimbabwean Journal of Botany*, Vol. 1 (Harare, ZW: Government Printers, 1960–1961), p. 5.

231 Raffensperger, *A Brief History of the Edinburgh School of Medicine*, p. 80.

232 Liebowitz, *The Physician and the Slave Trade: John Kirk, the Livingstone Expeditions, and the Crusade Against Slavery in East Africa*, pp. 11–12.

Endnotes

233 Wild, "Sir John Kirk," *Kirkia: Zimbabwean Journal of Botany*, Vol. 1, p. 5.

234 Tim Jeal, *Livingstone: Revised and Expanded Edition*, (New Haven, CT: Yale University Press, 2013).

235 Alan Moorehead, *The White Nile*, (London, UK: Harper & Row Publishers, 1960), p. 134.

236 Liebowitz, *The Physician and the Slave Trade: John Kirk, the Livingstone Expeditions, and the Crusade Against Slavery in East Africa*, pp. 13–20.

237 Moorehead, *The White Nile*, p. 115.

238 Ibid., p. 116.

239 Oliver Ransford, *David Livingstone: The Dark Interior*, (London, UK: John Murray, 1978), p. 4.

240 Jeal, *Livingstone: Revised and Expanded Edition*.

241 C. A. Baker, "The Development of the Administration to 1897," in *The Early History of Malawi*, edited by Bridglal Pachai (London, Longman, 1972), p. 324.

242 Richard B. Allen, *European Slave Trading in the Indian Ocean, 1500–1850*, (Athens, OH: Ohio University Press, 2014). pp. 295–299.

243 Kevin Shillington, *History of Africa. Revised second edition* (New York: Macmillan Publishers Limited, 2005), p. 301.

244 Liebowitz, *The Physician and the Slave Trade: John Kirk, the Livingstone Expeditions, and the Crusade Against Slavery in East Africa*, pp. 51–56.

245 Ibid., p. 20.

246 Ibid., pp. 60–62.

247 Tim Jeal, *Livingstone*, (New York, NY: G. p. Putnam's Sons, 1973).

248 Ed Wright, *Lost Explorers: Adventurers who Disappeared Off the Face of the Earth*, (Crow's Nest, AU: Allen & Unwin, 2008).

249 Ransford, *David Livingstone: The Dark Interior*, p. 218.

250 Francis Barrow Pearce, *Zanzibar: The Island Metropolis of Eastern Africa*, (New York, NY: Dutton and Company, 1920).

251 Sir Reginald Coupland, *The Exploitation of East Africa 1856–1890: The Slave Trade and the Scramble*, (Evanston, IL: Northwestern University Press, 1967), pp. 41–49.

252 Goring, *Scotland: The Autobiography*, pp. 258–259.

253 Claire Pettitt, *Dr Livingstone I Presume: Missionaries, Journalists, Explorers and Empire*, (London, UK: Profile Books, 2013), p. 156.

254 Moorehead, *The White Nile*, p. 118.

255 John Bierman, *Dark Safari: The Life Behind the Legend of Henry Morton Stanley*, (Austin, TX: University of Texas Press, 1990), pp. 4–27.

256 Moorehead, *The White Nile*, pp. 120–124.

257 Ibid., *The White Nile*, p. 128.

258 Liebowitz, *The Physician and the Slave Trade: John Kirk, the Livingstone Expeditions, and the Crusade Against Slavery in East Africa*, pp. 157–162.

259 Moorehead, *The White Nile*, pp. 131–134.

260 Coupland, *The Exploitation of East Africa 1856–1890: The Slave Trade and the Scramble*, p. 93.

261 Ibid., *The Exploitation of East Africa 1856–1890: The Slave Trade and the Scramble*, pp. 53–55.

262 Ibid., *The Exploitation of East Africa 1856–1890: The Slave Trade and the Scramble*, pp. 8–11.

263 Ibid., *The Exploitation of East Africa 1856–1890: The Slave Trade and the Scramble*, pp. 165–172.

264 Ibid., *The Exploitation of East Africa 1856–1890: The Slave Trade and the Scramble*, pp. 183–191.

265 Ibid., *The Exploitation of East Africa 1856–1890: The Slave Trade and the Scramble*, pp. 198–200.

266 Ibid., *The Exploitation of East Africa 1856–1890: The Slave Trade and the Scramble*, pp. 207–209.

267 Ibid., *The Exploitation of East Africa 1856–1890: The Slave Trade and the Scramble*, pp. 224–225.

268 Daniel Liebowitz, *England in the Nineteenth Century*, (New York, NY: Longmans, Green and Com., 1999), pp. 178–183.

269 Qman, *England in the Nineteenth Century*, p. 193.

270 Liebowitz, *The Physician and the Slave Trade: John Kirk, the Livingstone Expeditions, and the Crusade Against Slavery in East Africa*, pp. 239–241.

271 Ibid., pp. 251–257.

Chapter 8

272 John Rae, *The Arctic Journals of John Rae*, (Vancouver, CN: Touchwood Editions, 2012), p. 14.

273 Ken McGoogan, *Fatal Passage: The Story of John Rae, the Arctic Hero Time Forgot*, (New York, NY: Carroll & Graff Publishers, 2002), p. 10.

274 Rae, *The Arctic Journals of John Rae*, p. 14.

275 McGoogan, *Fatal Passage: The Story of John Rae, the Arctic Hero Time Forgot*, pp. 30–31.

276 Ibid., p. 36.

277 Ibid., pp. 15–19.

278 Ibid., pp. 40–42.

279 Rae, *The Arctic Journals of John Rae*, pp. 4–17.

280 Ibid., pp. 17–18.

281 McGoogan, *Fatal Passage: The Story of John Rae, the Arctic Hero Time Forgot*, p. 153.

282 Ibid., p. 186.

283 Ibid., pp. 188–190.

284 Rae, *The Arctic Journals of John Rae*, pp. 34–43.

285 McGoogan, *Fatal Passage: The Story of John Rae, the Arctic Hero Time Forgot*, pp. 193–197.

286 Ibid., pp. 201–208.

287 Ibid., pp. 209–212.

288 Ibid., pp. 289–290.

289 Anthony Kamm, *Scottish Explorers*, (Edinburgh, UK: National Museums of Scotland, 2013), pp. 5–7.

290 Ana Faguy, DNA test helps identify sailor from doomed Arctic expedition, BBC.com/news/article, 24 September 2024.

291 Rae, *The Arctic Journals of John Rae*, p. 6.

292 McGoogan, *Fatal Passage: The Story of John Rae, the Arctic Hero Time Forgot*, p. 305.

Chapter 9

293 Richard Corfield, *The Silent Landscape: The Scientific Voyage of HMS Challenger*, (Washington, DC: Joseph Henry

Endnotes

Press, 2003), p. 5.

294 Corfield, *The Silent Landscape: The Scientific Voyage of HMS Challenger*, p. 6.

295 "A History of the Study of Marine Biology," (Houston, TX: Marine BioConservation Society, Retrieved 17 May 2021), p. 1.

296 "A History of the Study of Marine Biology," p. 2.

297 Doug MacDougall, *Endless Novelties of Extraordinary Interest: The Voyage of H.M.S. Challenger and the Birth of Modern Oceanography*, (New Haven, CT: Yale University Press, 2019), pp. 20–21.

298 Corfield, *The Silent Landscape: The Scientific Voyage of HMS Challenger*, pp. 2–3.

299 Ibid., pp. 2–3.

300 Charles Wyville Thomson, *The Depths of the Sea: An Account of the General Results of the Dredging Cruises of H.M. SS. 'Porcupine' and 'Lightning' During the Summers of 1868, 1869, and 1870, Under the Scientific Direction of Dr. Carpenter, J. Gwyn Jeffreys, and Dr. Wyville Thomson*, (Charleston, SC: Nabu Press, 2014).

301 Corfield, *The Silent Landscape: The Scientific Voyage of HMS Challenger*, pp. 4–5.

302 Linklater, *The Voyage of the Challenger*, p. 14.

303 Corfield, *The Silent Landscape: The Scientific Voyage of HMS Challenger*, pp. 5–6.

304 Linklater, *The Voyage of the Challenger*, p. 15.

305 MacDougall, *Endless Novelties of Extraordinary Interest: The Voyage of HMS Challenger and the Birth of Modern Oceanography*, p. 20.

306 Ibid., pp. 8–10.

307 Corfield, *The Silent Landscape: The Scientific Voyage of HMS Challenger*, p. xiii.

308 Linklater, *The Voyage of the Challenger*, p. 270.

309 John Murray, *Report on the Scientific Results of the Voyage of HMS Challenger During the Years 1873–76 Under the Command of Captain George S. Nares and the Late Captain Frank Tourle Thomson*, (New York, NY: Franklin Classics, 2018).

310 MacDougall, *Endless Novelties of Extraordinary Interest: The Voyage of HMS Challenger and the Birth of Modern Oceanography*, p. xii.

Chapter 10

311 William S. Bruce, *Polar Exploration*, (New York, NY: Henry Holt and Company, 1911) pp. i–ii.

312 Special Collections, University of Edinburgh, *Attendance General* 1647/46/5.

313 Peter Speak, *William Speirs Bruce: Polar Explorer and Scottish Nationalist*, (Edinburgh, UK: NMS Publishing, 2003), pp. 31–32.

314 Isobel p. Williams and John Dudeney, *William Speirs Bruce: Forgotten Polar Hero*, (Gloucestershire, UK: Amberley Publishing, 2018), pp. 19–21.

315 Ibid., p. 24.

316 Speak, *William Speirs Bruce: Polar Explorer and Scottish Nationalist*, p. 34.

317 Ibid., p. 36.

318 Williams and Dudeney, *William Speirs Bruce: Forgotten Polar Hero*, pp. 29–30.

319 Ibid., pp. 33–34.

320 Speak, *William Speirs Bruce: Polar Explorer and Scottish Nationalist*, pp. 46–51.

321 Williams and Dudeney, *William Speirs Bruce: Forgotten Polar Hero*, pp. 39–48.

322 Speak, *William Speirs Bruce: Polar Explorer and Scottish Nationalist*, pp. 52–58.

323 Williams and Dudeney, *William Speirs Bruce: Forgotten Polar Hero*, pp. 51–62.

324 Ibid., p. 64.

325 Ibid., pp. 64–68.

326 Ibid., p. 69.

327 Ibid., pp. 78–79.

328 Ibid., pp. 66–72.

329 Speak, William *Speirs Bruce: Polar Explorer and Scottish Nationalist*, pp. 75–76.

330 Ibid., pp. 85–91.

331 Williams and Dudeney, *William Speirs Bruce: Forgotten Polar Hero*, p. 75.

332 Speak, *William Speirs Bruce: Polar Explorer and Scottish Nationalist*, pp. 90–91.

333 Williams and Dudeney, *William Speirs Bruce: Forgotten Polar Hero*, pp. 99–100.

334 Ibid., pp. 115–121.

335 Speak, *William Speirs Bruce: Polar Explorer and Scottish Nationalist*, pp. 100–101.

336 Ibid., pp. 133–134.

337 Ibid., pp. 12–13.

Chapter 11

338 Timothy H. Parsons, *The British Imperial Century, 1815–1914: A World History Perspective*, (Lanham, MD: Rowman & Littlefield Publishers, 2019), 13–25.

339 Qman, *England in the Nineteenth Century*, p. 212.

340 John McLeod, *The Routledge Companion to Postcolonial Studies*, (New York, NY: Routledge Publishing, 2007).

341 Brown, *Sightseers and Scholars: Scientific Travelers in the Golden Age of Natural History*, p. 13.

342 Allan and Forsyth, *Common Cause: Commonwealth Scots and the Great War*, p. 2.